METHODS IN MOLECULAR BIOLOGY

Series Editor
John M. Walker
School of Life and Medical Sciences
University of Hertfordshire
Hatfield, Hertfordshire, AL10 9AB, UK

For further volumes:
http://www.springer.com/series/7651

Protein Function Prediction

Methods and Protocols

Edited by

Daisuke Kihara

Department of Biological Sciences and Computer Science
Purdue University
West Lafayette, Indiana, USA

Humana Press

Editor
Daisuke Kihara
Department of Biological Sciences and Computer Science
Purdue University
West Lafayette, Indiana, USA

ISSN 1064-3745 ISSN 1940-6029 (electronic)
Methods in Molecular Biology
ISBN 978-1-4939-7013-1 ISBN 978-1-4939-7015-5 (eBook)
DOI 10.1007/978-1-4939-7015-5

Library of Congress Control Number: 2017937538

Printed on acid-free paper

This Humana Press imprint is published by Springer Nature
The registered company is Springer Science+Business Media LLC
The registered company address is: 233 Spring Street, New York, NY 10013, U.S.A.

Preface

Knowing the function of a protein and understanding how it is carried out are the ultimate goals of molecular biology and biochemistry. From the early stage of bioinformatics in the 1980s, the development of computational tools to aid in elucidating protein function was a major focus of the field. Numerous methods have been developed since then. Computationally, protein function can be predicted through similarity searches because similarity implies homology from an evolutionary standpoint, and also because it indicates that the proteins have the same physical structures where the function takes place. Thus, based on this similarity principle, methods were developed to compare global or local sequences and the structures of proteins. Databases were also developed, which organize function information of proteins and serve as references to be queried against. In this book, well-established sequence- and structure-based tools and databases are introduced, which are very useful for biology labs. In addition, this book introduces software which addresses function beyond its conventional meaning, reflecting the diversity of the current active research field.

This book begins by introducing two sequence-based function prediction methods, PFP and ESG, in Chapter 1. The chapter also describes a web server, NaviGO, which can analyze Gene Ontology annotations. Then, Chapters 2, 3, and 4 discuss tools suitable for the functional analysis of metagenomics data. The tools in these three chapters are based on sequence database searches faster than conventional homology search methods, a necessity when processing the large amounts of sequence data which typify metagenome sequences. Chapter 2 introduces GhostX, which uses a suffix array for fast sequence comparison. Fun4Me in Chapter 3 is a pipeline that combines protein coding gene detection in query sequences and a fast sequence database search utilizing a hashing technique. SUPER-FOCUS in Chapter 4 combines fast search algorithms with preclustered reference sequence databases. In Chapter 5, we have MPFit, a program that detects when query proteins are moonlighting proteins, i.e., a protein with dual functions.

The next chapter (Chapter 6) describes SignalP, a well-established web server that predicts subcellular localization by recognizing a signal peptide in a query sequence. Subcellular localization is one of the three functional categories in the Gene Ontology (Cellular Component), and it can be a clue for other biological functions of a protein since localization and biological function are closely correlated.

The following four chapters deal with protein structures. ProFunc in Chapter 7 is a popular web server that performs multiple different analyses on a query protein structure, including global and local structure matching to known proteins. Chapter 8 describes G-LoSA, which finds ligand binding sites similar to a query binding site within a reference database. eMatchSite, the following chapter (Chapter 9), aligns two ligand binding sites to quantify similarities between them. In Chapter 10, WATsite2.0 is introduced, which predicts bound water molecules in a ligand binding site. Water molecules bound to proteins mediate ligand-protein interactions and are thus important in protein function.

The subsequent five chapters cover resources that address protein function through pathways, networks, and genomes. Chapter 11 discusses recent updates of KEGG, focusing on enzymes and pathways. KEGG is one of the most comprehensive databases of pathways, genomes, and other biomolecules and is a fundamental resource for understanding protein function at a systems level. Chapter 12 is about the Microbial Genome Database, a valuable resource to perform comparative genomics. The *Saccharomyces* Genome Database (SGD) is described in Chapter 13. *S. cerevisiae* is one of the most extensively studied organisms. SGD has long served as a reliable source for protein function and other resources, including gene expression and phenotypes, in *S. cerevisiae*. Chapter 14 introduces MouseNet, which predicts gene function in mice from a gene expression network. FANTOM5 in Chapter 15 is a database of human and mouse genomes. Transcription start sites and promoter activities of various cells can be browsed and searched. The last chapter (Chapter 16) introduces Spatiocyte, a software for simulating the diffusion and localization of proteins in a cell. Results from the simulation, i.e., a phenotype, can be compared against microscope observations. Proteins exhibit their function through dynamic interactions in a cell environment. Thus, ultimately functions must be considered in a dynamic system, which this software aims to do.

I hope readers enjoy this book as a practical guide for using bioinformatics tools related to protein function prediction. Moreover, I also hope that this compilation itself exhibits a snapshot of the current research field and our understanding of the concept of protein function, while indicating the future direction of the field.

Editing of this book was greatly aided by Mr. Joshua McGraw, Ms. Sarah Rodenbeck, Ms. Lenna X. Peterson, and Mr. Charles Christoffer of my research group. I would like to conclude this preface by recognizing and acknowledging their help as a happy memory of my research activities.

West Lafayette, IN, USA *Daisuke Kihara*

Contents

List of Contributors

Imad Abugessaisa • *Division of Genomics Technologies, RIKEN Center for Life Science Technologies, Yokohama, Kanagawa, Japan*

Yutaka Akiyama • *Department of Computer Science, School of Computing, Tokyo Institute of Technology, Tokyo, Japan; Education Academy of Computational Life Sciences (ACLS), Tokyo Institute of Technology, Yokohama, Japan; Department of Computer Science, Graduate School of Information Science and Engineering, Tokyo Institute of Technology, Tokyo, Japan*

Satya N.V. Arjunan • *Laboratory for Biochemical Simulation, RIKEN Quantitative Biology Center, Suita, Osaka, Japan*

Michal Brylinski • *Department of Biological Sciences, Louisiana State University, Baton Rouge, LA, USA; Center for Computation & Technology, Louisiana State University, Baton Rouge, LA, USA*

Piero Carninci • *Division of Genomics Technologies, RIKEN Center for Life Science Technologies, Yokohama, Kanagawa, Japan*

Robert A. Edwards • *Computational Science Research Center, San Diego State University, San Diego, CA, USA; Department of Biology, San Diego State University, San Diego, CA, USA; Department of Computer Science, San Diego State University, San Diego, CA, USA*

Bingjie Hu • *Department of Medicinal Chemistry and Molecular Pharmacology, College of Pharmacy, Purdue University, West Lafayette, IN, USA; Computational ADME, Drug Disposition, Lilly Research Laboratories, Eli Lilly and Company, Indianapolis, IN, USA*

Wonpil Im • *Department of Biological Sciences and Bioengineering Program, Lehigh University, Bethlehem, PA, USA*

Takashi Ishida • *Department of Computer Science, School of Computing, Tokyo Institute of Technology, Tokyo, Japan; Education Academy of Computational Life Sciences (ACLS), Tokyo Institute of Technology, Yokohama, Japan; Department of Computer Science, Graduate School of Information Science and Engineering, Tokyo Institute of Technology, Tokyo, Japan*

Takeya Kasukawa • *Division of Genomics Technologies, RIKEN Center for Life Science Technologies, Yokohama, Kanagawa, Japan*

Masanori Kakuta • *Department of Computer Science, Graduate School of Information Science and Engineering, Tokyo Institute of Technology, Tokyo, Japan*

Minoru Kanehisa • *Institute for Chemical Research, Kyoto University, Uji, Kyoto, Japan*

Ishita Khan • *Department of Computer Science, Purdue University, West Lafayette, IN, USA*

Daisuke Kihara • *Department of Biological Sciences and Computer Science, Purdue University, West Lafayette, IN, USA*

Eiru Kim • *Department of Biotechnology, College of Life Science and Biotechnology, Yonsei University, Seoul, Korea*

Roman A. Laskowski • *European Bioinformatics Institute, Hinxton, Cambridge, UK*

Hui Sun Lee • *Department of Biological Sciences and Bioengineering Program, Lehigh University, Bethlehem, PA, USA*

Insuk Lee • *Department of Biotechnology, College of Life Science and Biotechnology, Yonsei University, Seoul, Korea*

Markus A. Lill • *Department of Medicinal Chemistry and Molecular Pharmacology, College of Pharmacy, Purdue University, West Lafayette, IN, USA*
Joshua McGraw • *Department of Biological Sciences, Purdue University, West Lafayette, IN, USA*
Fabyano A.C. Lopes • *Cellular Biology Department, Universidade de Brasília (UnB), Brasília, DF, Brazil*
Henrik Nielsen • *Department of Bio and Health Informatics, Technical University of Denmark, Lyngby, Denmark*
Shuhei Noguchi • *Division of Genomics Technologies, RIKEN Center for Life Science Technologies, Yokohama, Kanagawa, Japan*
Masahito Ohue • *Department of Computer Science, Graduate School of Information Science and Engineering, Tokyo Institute of Technology, Tokyo, Japan; Department of Computer Science, School of Computing, Tokyo Institute of Technology, Tokyo, Japan*
Fatemeh Sharifi • *School of Informatics and Computing, Indiana University, Bloomington, IN, USA*
Genivaldo Gueiros Z. Silva • *Computational Science Research Center, San Diego State University, San Diego, CA, USA*
Shuji Suzuki • *Department of Computer Science, Graduate School of Information Science and Engineering, Tokyo Institute of Technology, Tokyo, Japan; Education Academy of Computational Life Sciences (ACLS), Tokyo Institute of Technology, Yokohama, Japan*
Koichi Takahashi • *Laboratory for Biochemical Simulation, RIKEN Quantitative Biology Center, Suita, Osaka, Japan*
Ikuo Uchiyama • *Laboratory of Genome Informatics, National Institute for Basic Biology, National Institutes of Natural Sciences, Okazaki, Aichi, Japan*
Qing Wei • *Department of Computer Science, Purdue University, West Lafayette, IN, USA*
Edith D. Wong • *Department of Genetics, Stanford University, Stanford, CA, USA*
Ying Yang • *Department of Medicinal Chemistry and Molecular Pharmacology, College of Pharmacy, Purdue University, West Lafayette, IN, USA*
Yuzhen Ye • *School of Informatics and Computing, Indiana University, Bloomington, IN, USA*

Chapter 1

Using PFP and ESG Protein Function Prediction Web Servers

Qing Wei, Joshua McGraw, Ishita Khan, and Daisuke Kihara

Abstract

Elucidating biological function of proteins is a fundamental problem in molecular biology and bioinformatics. Conventionally, protein function is annotated based on homology using sequence similarity search tools such as BLAST and FASTA. These methods perform well when obvious homologs exist for a query sequence; however, they will not provide any functional information otherwise. As a result, the functions of many genes in newly sequenced genomes are left unknown, which await functional interpretation. Here, we introduce two webservers for function prediction methods, which effectively use distantly related sequences to improve function annotation coverage and accuracy: Protein Function Prediction (PFP) and Extended Similarity Group (ESG). These two methods have been tested extensively in various benchmark studies and ranked among the top in community-based assessments for computational function annotation, including Critical Assessment of Function Annotation (CAFA) in 2010–2011 (CAFA1) and 2013–2014 (CAFA2). Both servers are equipped with user-friendly visualizations of predicted GO terms, which provide intuitive illustrations of relationships of predicted GO terms. In addition to PFP and ESG, we also introduce NaviGO, a server for the interactive analysis of GO annotations of proteins. All the servers are available at http://kiharalab.org/software.php.

Keywords Protein function prediction, Genome annotation, BLAST, Gene Ontology, Automated function prediction, Sequence analysis

1 Introduction

Functional interpretation of novel proteins is a central problem in molecular biology and bioinformatics. As genome sequencing and proteomic technologies advance at a striking pace, an overwhelming amount of sequence data awaits to be analyzed and assigned with functional interpretations. Since performing biological experiment for such purposes does not scale up in terms of time, effort and expense, automatic function prediction (AFP) methods have been pursued and have become one of the important problems in bioinformatics. There are many AFP algorithms developed in the past years in order to achieve accurate annotation and wider coverage to replace the conventional function prediction methods which use homology as the source of information [1, 2]. A review by

Daisuke Kihara (ed.), *Protein Function Prediction: Methods and Protocols*, Methods in Molecular Biology, vol. 1611,
DOI 10.1007/978-1-4939-7015-5_1,

Hawkins & Kihara summarizes several categories of AFP methods beyond traditional sequence similarity, which leverage sequence, structural, genomic, cellular and metabolic context-based information [3]. A review by Sael et al. [4] focuses on AFP methods for non-homologous proteins in the sequence and structure-based categories.

For the advancement of such computational techniques, it is very important that there are community-wide efforts for objective evaluation of prediction accuracy. Among several efforts carried out in the protein function prediction community in the past, a recent notable one is CAFA (Critical Assessment of Function Annotation) [5]. The first round of CAFA was held in 2010–2011 [5], and the second round, CAFA2, was held in 2013–2014 [6]. CAFA3 is planned in 2016–2017.

Here, we introduce two publicly available webservers for function prediction methods: Protein Function Prediction (PFP) [7, 8] and Extended Similarity Group (ESG) [9]. Both webservers take a list of query sequences and output a list of predicted Gene Ontology (GO) terms [10, 11]. The servers have been maintained over years and extensively benchmarked in the past [12, 13]. In both CAFA1 and CAFA2, PFP and ESG were ranked among the top function prediction methods. In the CAFA1 experiment, ESG was ranked fourth in the molecular function (MF) GO category among 54 participating groups [5], while PFP did well in all the three categories in CAFA 2 [6]. In an earlier community-based assessment, the function prediction category of (CASP) held in 2006, PFP was ranked the top [14].

PFP and ESG were designed to achieve complementary goals: PFP is for achieving a large prediction coverage by retrieving annotations widely including from weakly similar sequences. On the other hand, ESG is for improving specificity by accumulating contribution of consistently predicted GO terms in an iterative search. The interactive webserver of PFP and ESG [15] is developed to assist in the sequence-based function prediction and to enhance the understanding of predicted functions by an effective visualization of the predictions in a hierarchical GO topology. In addition, we also describe NaviGO, a newly developed web-based tool for interactive analysis of GO term annotations of proteins. All the servers are available at http://kiharalab.org/software.php.

2 Function Prediction Algorithms in PFP and ESG

In this section, we briefly explain the main idea of PFP and ESG algorithms. For more details, please refer to the original papers [7–9].

2.1 The PFP Algorithm

The PFP algorithm uses PSI-BLAST [1] to obtain sequence hits for a target sequence and computes the score for GO term f_a as follows:

$$s(f_a) = \sum_{i=1}^{N} \sum_{j=1}^{N\text{func}(i)} \left((-\log(E - \text{value}(i)) + b) P\left(f_a | f_j\right) \right) \tag{1}$$

where N is the number of sequence hits considered in the PSI-BLAST hits; Nfunc(i) is the number of GO annotations for the sequence hit i; E-value(i) is the PSI-BLAST E-value for the sequence hit i; f_j is the jth annotation of the sequence hit i; and constant b takes value 2 ($= \log_{10} 125$) to keep the score positive when retrieved sequences up to an E-value of 125 are used. The conditional probabilities $P(f_a|f_j)$ are used to consider co-occurrence of GO terms in a single sequence annotation, which are computed as the ratio of the number of proteins co-annotated with GO terms f_a and f_j as compared with ones annotated only with the term f_j. To take into account the hierarchical structure of GO, PFP transfers the raw score to the parental terms by computing the proportion of proteins annotated with f_a relative to all proteins that belong to the parental GO term in the database. The score of a GO term computed as the sum of the directly computed score by Eq. 1 and the ones from the parental propagation is called the raw score.

Compared to the conventional usage of PSI-BLAST that uses a strict E-value cutoff, e.g., 0.001, for transferring function annotations, the characteristic of PFP is that it collects GO annotations even from very weakly sequences up to an E-value of 125. Individual weakly similar sequences do not contribute much to a raw score, but a GO term can accumulate a substantially large score and be predicted with confidence if the GO term appears in many sequences.

2.2 The ESG Algorithm

ESG recursively performs PSI-BLAST searches from sequence hits obtained in the initial search from the query sequence Q, which will retrieve N sequence hits (N is "the number of hits per stage" parameter in the ESG input page as shown in the next section), S_1, S_2,...S_N, each with E-value E_1, E_2,...E_N, respectively. Each sequence hit in a search is assigned a weight W_i that is computed as the proportion of the $-\log$(E-value) of the sequence relative to the sum of the $-\log$(E-value) from all the sequence hits considered in the search of the same level:

$$W_i = \frac{-\log(E_i) + b}{\sum_{j=1}^{N} \{-\log(E_j) + b\}} \tag{2}$$

where score $-\log(E_i)$ is shifted by a constant value b, which makes the score a nonnegative value. This weight is assigned for GO terms annotating the sequence hit and the probability of the GO term f_a annotating the query sequence Q is defined as the sum of weights of f_a that come from sequences annotated with f_a:

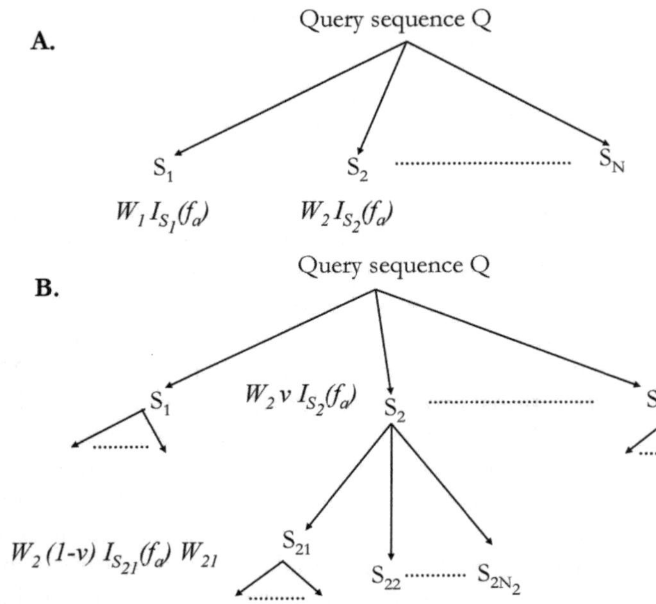

Fig. 1 Computing the ESG score. (**a**) For a single-layer search, a score of a function f_a is computed as a sum of the weight of sequences that have f_a in their GO annotation. (**b**) When a two-layer search is performed, a score comes from a weighted combination of the second level search and the first level search. This figure is adopted from the original paper of ESG (Chitale, Hawkins, Park, & Kihara, Bioinformatics, 25: 1739–1745, 2009) with permission from the publisher

$$P_{\mathrm{Q}}^{\mathrm{d}}(f_{\mathrm{a}}) = \sum_{i=1}^{N} W_i \cdot I_{S_i}(f_{\mathrm{a}}) \tag{3}$$

The function I indicates whether the given sequence S_i has annotation f_a:

$$I_{S_i}(f_{\mathrm{a}}) = \begin{cases} 1 \text{ if } S_i \text{ has } f_{\mathrm{a}} \text{ annotation} \\ 0 \text{ otherwise} \end{cases} \tag{4}$$

The index d on the left side of Eq. 3 indicates that function information comes from direct annotations to sequences. Additionally, multilevel exploration ("the number of stages" parameter in the ESG input page) of the sequence-similarity space (PSI-BLAST) shown in Fig. 1 is performed around the target protein by sharing the weights between levels using a weight parameter v. In the second round, each of the sequences $S_1, S_2, \ldots S_N$ retrieved in the first round is in turn used as a query. Suppose sequence S_i obtains N_i sequences by a PSI-BLAST run, each referred to as S_{ij}. The weights for S_{ij}, W_{ij} can be computed in a similar manner to Eq. 2. Combining the two levels of searches:

$$P_Q^d(f_a) = \sum_{i=1}^{N} W_i \cdot P_{S_i}^d(f_a) \quad (5)$$

$$P_S^d(f_a) = v \cdot I_{S_i}(f_a) + (1 - v) \cdot \sum_{j=1}^{N_i} W_{ij} \cdot I_{S_{ij}}(f_a) \quad (6)$$

Equation 5 is a variation of Eq. 3, representing that the score of a GO term f_a for the query Q is contributed by sequences retrieved at the first level (S_1 to S_N). The weights for GO terms found in the second level search are computed similarly, where Eq. 2 defines the weight W_i. Eq. 6 defines the score for f_a for sequence S_i as a combination of $I_{S_i}(f_a)$, which is sequence S_i's annotation, and the second level search. The first and the second terms are weighted by a factor v. Moreover, the equations can be recursively extended to multiple levels of searches to explore broader space around the query sequence. The score for each GO term ranges from 0.0 to 1.0.

ESG predicts a GO term with a high score if it appears many times consistently in the multiple searches including the initial search and the second level searches. In general, the number of GO terms predicted by ESG is smaller (5–10 GO terms) than PFP (often over 50 terms), and terms predicted by high scores by ESG are usually highly accurate.

3 Input and Output of the Servers

3.1 Query Input Page of PFP and ESG

In Subheading 3, we explain how to use the webservers with an example. PFP is available at http://kiharalab.org/pfp.php and ESG is at http://kiharalab.org/esg.php. Query sequences can be submitted to both PFP and/or ESG from the combined submission page at http://kiharalab.org/web/pfp_esg.php. Please also refer to a detailed instruction at http://kiharalab.org/web/pfp_tutorial.php and http://kiharalab.org/web/esg_tutorial.php for PFP and ESG, respectively. Both the servers may be used without making an account; however, users are encouraged to create their account on the servers. With an account, users may automatically keep and refer to prediction results that have been processed earlier.

PFP and ESG accept query inputs of FASTA formatted protein sequences. Users may submit sequences separated by line breaks in the text box titled "Enter Query Sequence(s)" or upload a FASTA file containing multiple sequences (Fig. 2). To view a sample of the format, users may click on "Load Sample" to fill the field with an example sequence. Selecting "Clear" will remove all inputs sequences including uploaded files. Currently, up to 100 sequences

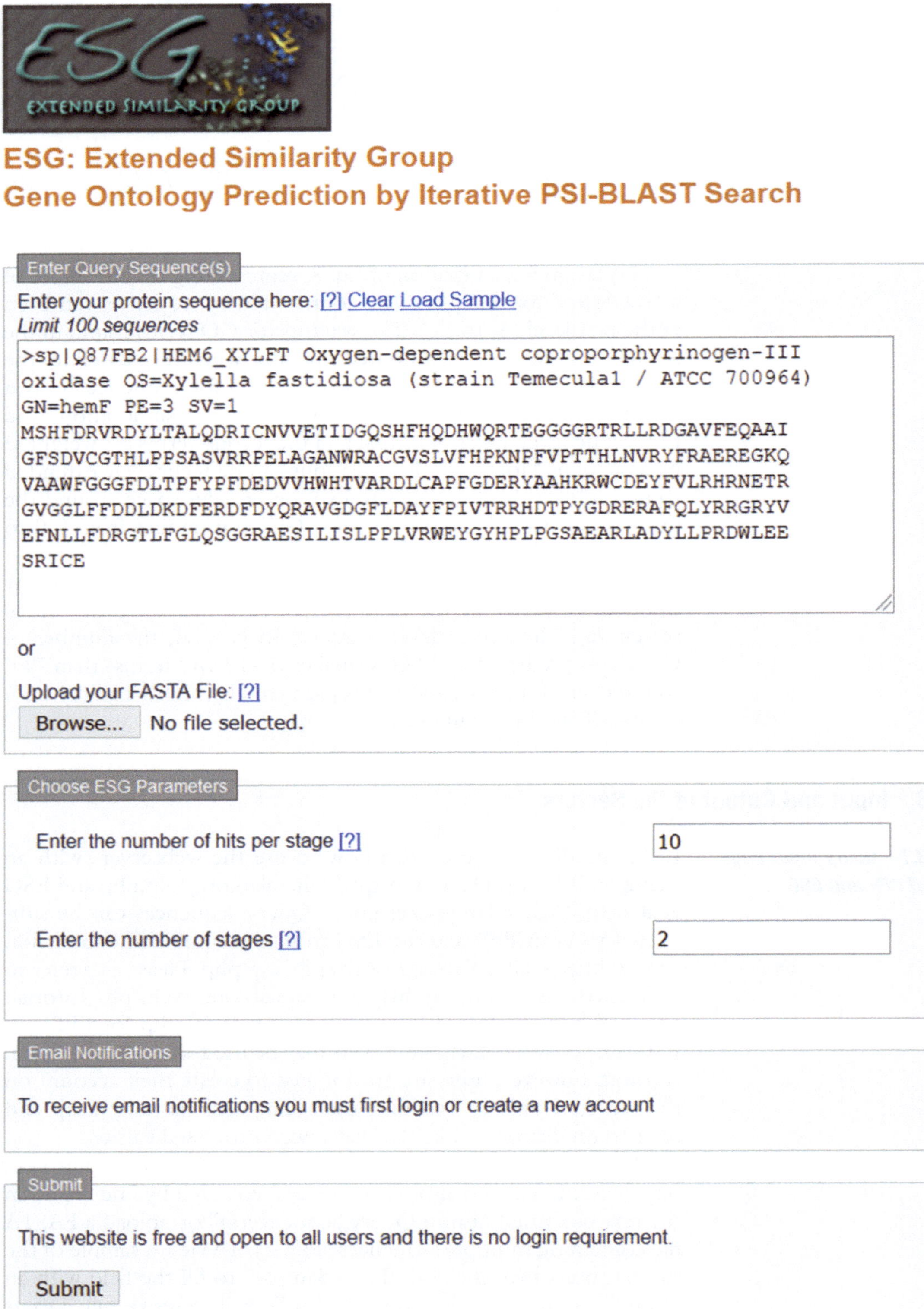

Fig. 2 Query input page of ESG. Query sequences can be pasted in the submission window or a sequence file can be uploaded. The query page of PFP is essentially the same, except that it does not have the number of hits and the number of stages parameters

may be submitted at a time to avoid overloading the computer server by the job queue.

For ESG, there are two more parameters that must be entered: "Number of hits" and "Number of stages." "Number of hits" indicates the number of PSI-BLAST hits to be considered at each level of ESG. The default value of this parameter is set to 10 in our web server. "Number of stages" indicates the level of searches to be performed by ESG. The default value for this parameter is chosen as 2. We recommend not changing the "Number of stages" parameter to a larger value as the computational time will suffer exponentially and we did not observe an improvement during benchmark in the original paper [9]. As for the "Number of hits" parameter, it can be increased if a prediction result by the default value is not satisfactory. For example, we used 50 for this value since it performed well during the benchmark [9]. However, if the parameter value is increased from 10 to 50, it requires roughly five times more computational time (with the two-stage setting).

3.2 Output Page with Case Studies

After selecting the submit button at the bottom section of the page, users will be directed to the job page displaying the status of that job. The job will be queued and assigned CPU time when available. You may refresh the page manually to check the status. Average computational time for PFP and ESG is 40.1 s and 7.5 min [15], respectively. When the job is completed, clicking on the job ID will display the predicted GO terms for the query sequences. Below we explain in detail how the results are presented.

3.2.1 PFP Output Page

The PFP results page shows the input sequences at the top section followed by the predicted terms for each GO category (Molecular Function (MF), Biological Process (BP), and Cellular Component (CC)), which have confidence greater than 5% of score of the top hit (Fig. 3). The results page also provides a link to the results in the XML format, which users may download for further processing. Selecting "Visualization of Predicted GO terms" will allow users to view the predicted terms in an interactive GO hierarchy. This tool allows users to pan and zoom through sub-nodes of related branches and is color mapped based on their assigned probability. Alternatively, users may select to color the nodes based on the number of child nodes under predicted terms. There are three different layouts users may choose (tree, radial, and circle) for visualizing the GO hierarchy as well as configurable layouts and interactive nodes in the Cytoscape [16] (Fig. 4).

Three links are provided below the visualization redirect links, which allow users to download static images of the GO hierarchy visualization. Selecting to download the image will render the SVG image and generate a figure. At the top of each static image is also a link to download the PNG image file. Users may also save the SVG image by bookmarking the static page for future reference.

List of Predicted GO Terms (the raw score can range 0 to over 20K)

Very high confidence : > 20K
High confidence : > 10K
Moderate confidence : > 500
Low confidence: >= 100 (But worthwhile to examine)
Below low confidence: < 100

Molecular Function Terms

PFP Score	Term	Description
36238.68	GO:0004109 [+]	coproporphyrinogen oxidase activity
36184.18	GO:0042803 [+]	protein homodimerization activity
30816.91	GO:0042802	identical protein binding
12008.03	GO:0046983	protein dimerization activity
10454.21	GO:0016634	oxidoreductase activity, acting on the CH-CH group of donors, oxygen as acceptor
5 Predictions 3 Predictions > 20K; 2 Predictions > 10K; 0 Predictions > 500; 0 Predictions >= 100		

Biological Process Terms

PFP Score	Term	Description
68236.29	GO:0006779 [+]	porphyrin-containing compound biosynthetic process
67396.63	GO:0033014	tetrapyrrole biosynthetic process
66437.40	GO:0006778	porphyrin-containing compound metabolic process
65630.69	GO:0033013	tetrapyrrole metabolic process
46970.40	GO:0042168	heme metabolic process
36243.03	GO:0055114 [+]	oxidation-reduction process
36015.28	GO:0006782 [+]	protoporphyrinogen IX biosynthetic process
35920.00	GO:0046501	protoporphyrinogen IX metabolic process
29808.98	GO:0046148	pigment biosynthetic process
29014.25	GO:0042440	pigment metabolic process
24438.22	GO:0006783 [+]	heme biosynthetic process
15670.98	GO:0051188	cofactor biosynthetic process
11611.37	GO:0051186	cofactor metabolic process
7816.74	GO:0008152 [+]	metabolic process
7519.68	GO:0018130	heterocycle biosynthetic process
15 Predictions 11 Predictions > 20K; 2 Predictions > 10K; 2 Predictions > 500; 0 Predictions >= 100		

Cellular Component Terms

PFP Score	Term	Description
34761.10	GO:0005737 [+]	cytoplasm
31279.85	GO:0044424	intracellular part
31040.66	GO:0005622 [+]	intracellular
25705.33	GO:0044464	cell part
25704.85	GO:0005623	cell
5 Predictions 5 Predictions > 20K; 0 Predictions > 10K; 0 Predictions > 500; 0 Predictions >= 100		

Fig. 3 An example of predicted GO terms by PFP is shown in the PFP output page. The query used is hemF, oxygen-dependent coproporphyrinogen-III oxidase (UniProt ID: Q87FB2). Each category of GO terms is separated by Molecular Function (MF), Biological Process (BP), and Cellular Component (CC). Prediction confidence is annotated by the color of the PFP Score, whereas *red* is very high confidence (>20 K) and *blue* is low confidence (100–500)

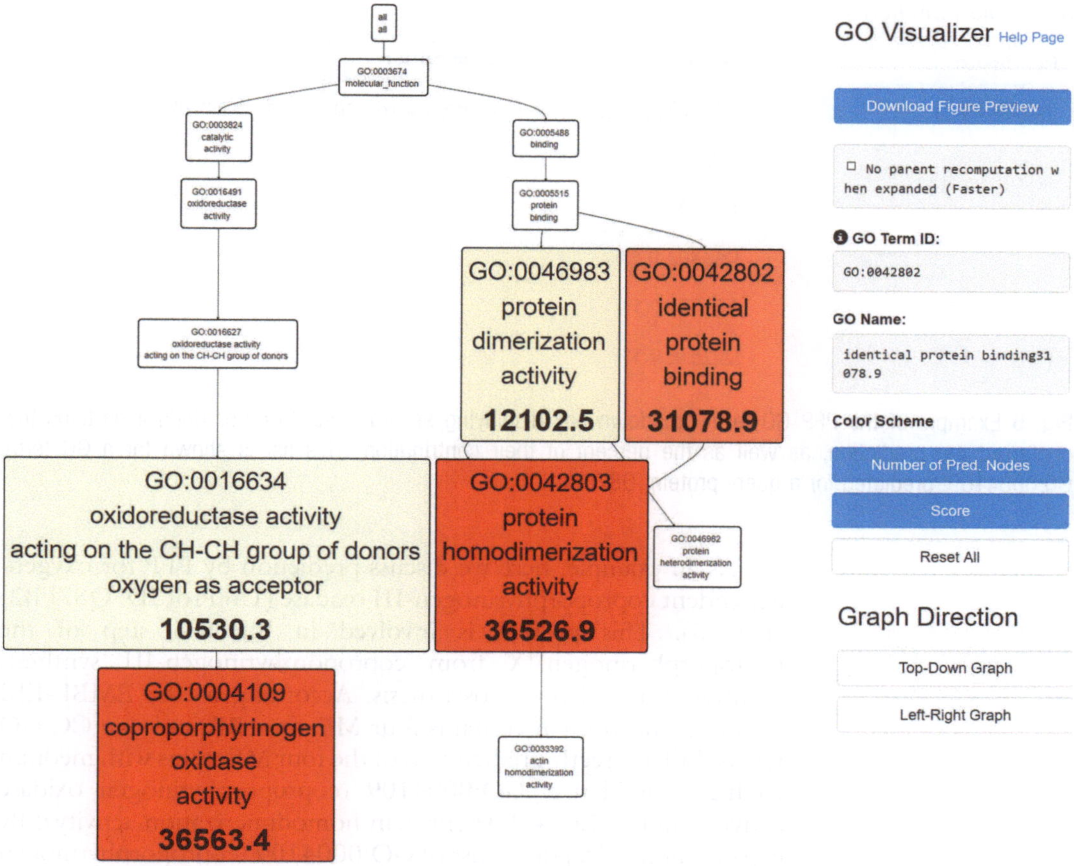

Fig. 4 Cytoscape output demonstrating a hierarchical Tree Layout of the PFP prediction. Each node represents a predicted GO term. *Red shades* in this figure indicate the prediction confidence

At the bottom section of the output page, the predicted results are categorized by MF, BP, and CC GO terms including the confidence, term ID, and term description. GO terms are colored as red, orange, green, and black, whereas red indicates high confidence of prediction (>70%) and black represents a low confidence (<30%). PFP allows users to trace the origin of the predicted GO terms through a dropdown list. Since the PFP algorithm often retrieves GO annotations from distantly related sequences that may not be obvious homologs, this tool provides useful insights as to how predictions are computed and the function of the query sequence. For each predicted GO term, clicking the [+] sign will open a dropdown list of sequence IDs which contributed toward the prediction. The contribution of each sequence is shown as the percentage of the score that originates from similar sequences (Fig. 5).

Molecular Function Terms

PFP Score	Term	Description
36238.68	GO:0004109 [-]	coproporphyrinogen oxidase activity
	B7MHT7 0.5%	
	B7MY86 0.5%	
	C4ZX09 0.5%	
	A8A2T4 0.5%	
	Q0TF33 0.5%	

Fig. 5 Example of the PFP GO term dropdown box displaying several links to other UniProt proteins that conferred the prediction, as well as the percent of their contribution. This list is shown for a GO term, GO:0004109, predicted for a query protein, Q87FB2

As an example, here we discuss prediction by PFP for oxygen-dependent coproporphyrinogen-III oxidase (UniProt ID: Q87FB2) (Fig. 3). This protein is involved in the first step of the protoporphyrinogen-IX from coproporphyrinogen-III synthesis pathway during heme biosynthesis. According to the EMBL-EBI database, this protein contains four MF, four BP, and one CC GO terms. PFP correctly predicts two of the four MF terms with medium to high confidence: GO:0004109 (coproporphyrinogen oxidase activity) and GO:0042803 (protein homodimerization activity). By expanding the dropdown list of GO:0004109 (coproporphyrinogen oxidase activity), we can trace the proteins that confer this prediction (Fig. 5). Proteins include hemF of *Escherichia coli O6:K15:H31* (UniProt ID: Q0TF33) (the protein in the bottom of Fig. 5) in the list serve to catalyze the aerobic oxidative decarboxylation of propionate groups of rings A and B of coproporphyrinogen-III to yield the vinyl groups in protoporphyrinogen-IX, and thus have the annotation of GO:0004109.

All four BP terms are predicted by PFP with very high confidence, which are GO:0006779 (porphyrin-containing compound biosynthetic process), GO:0006782 (protoporphyrinogen IX biosynthetic process), GO:0006783 (heme biosynthetic process), and GO:0055114 (oxidation-reduction process). Expanding the dropdown of GO:0006779 (porphyrin-containing compound biosynthetic process) reveals other hemF proteins such as (UniProtID: B7M6U5) of *Escherichia coli O8 (strain IAI1)* which support this prediction. PFP also correctly predicts the only CC term, GO:0005737 (cytoplasm), with very high confidence (Fig. 3).

3.2.2 ESG Output Page

Both ESG and PFP display identically formatted results page. To understand ESG's output page, refer to Subheading 3.2.1. PFP Output page.

3.3 GO Term Analysis Using NaviGO

In the last section, we introduce NaviGO, a recently developed web-based tool for Gene Ontology visualization and similarity quantification, which is useful for understanding the relationships between predicted GO terms. It is accessible at http://kiharalab.org/web/navigo.

To enable a quantitative analysis of GO terms and gene functions from various aspects, on NaviGO, users can compute similarity of GO terms using six different scoring schemes that incorporate a variety of information ranging from GO topological structure, contextual association, and GO annotation frequency. There are four major functionalities, which are accessible through tabs on the top bar of the web site, i.e., GO Parents, GO Set, GO Enrichment, and Protein Set.

In the GO Parents page, users are able to retrieve parental GO terms in the GO hierarchy (Directed Acyclic Graph, DAG) for a list of query GO terms. It uses a lite version of GO Visualizer [15] to help users understand relationships of GO terms topologically in the GO DAG. Results are rendered in an interactive DAG where query GO terms are circled with bold black outlines. Additionally, parental GO terms will be listed in the text area below the visualization.

In the GO Set page, the tool will compute pairwise GO similarity scores for a list of input GO terms and output them as three formats (Fig. 6): a table, a network graph, and a bubble chart.

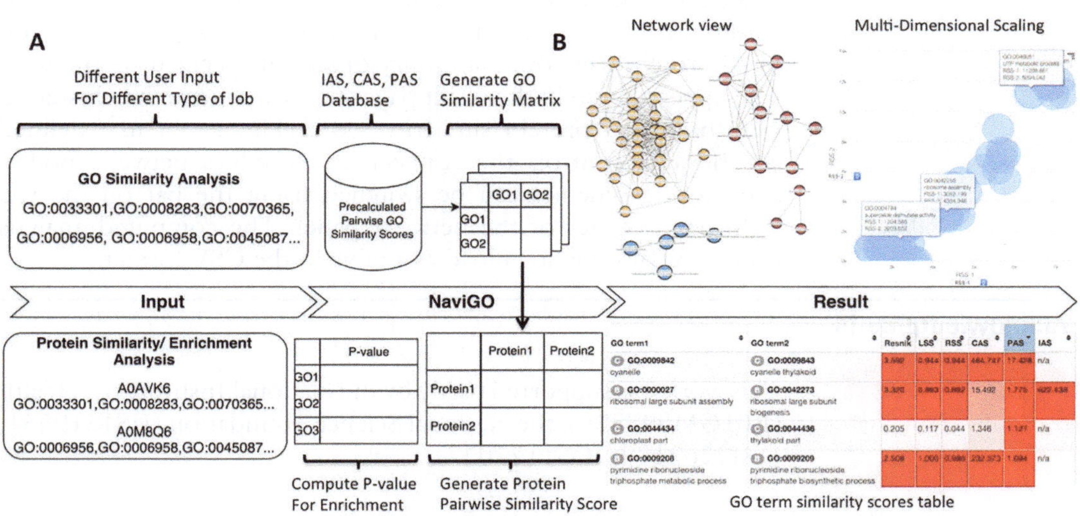

Fig. 6 (**a**) Workflow for NaviGO. Two types of input data are accepted, a set of GO terms or a set of genes with GO annotations. Similarity of GO terms is computed with six different GO scores including IAS, CAS, and PAS. If input data is a list of genes, then pairwise similarity scores for each pair of genes are computed. If GO enrichment analysis is selected, statistical significance of enrichment of GO terms is computed. (**b**), Presentation of results in NaviGO. Results are provided by a network view where similar GO terms or genes are connected; and in a bubble chart where similarity of GO terms is shown in a 2D plot of multi-dimensional scaling, or in a tabulated fashion, where significance of score similarity is indicated by a color scale

In the result table, Resnik's Similarity, Lin's Similarity, Relevance Similarity [17], Co-occurrence Association Score, PubMed Association Score [18], and Interaction Association Score [19] of pairs of input GO terms are colored based on score cutoffs. Table columns are sortable by clicking on score names at top row of the table. Common parents between a pair of GO terms are shown in the last column as well as a link to the interactive visualization, which illustrates parental GO terms in the GO DAG. In the network graph format, we showed an interactive network that summarizes the GO similarity as clusters where nodes are GO terms and edges indicate similarity score above a user-defined cutoff. The bubble chart format uses multidimensional scaling [20] to map the similarity into 2D coordinates and the user is able to choose the scoring schemes for either X or Y coordinates.

In the GO Enrichment tab, NaviGO will take the NCBI taxonomy ID of the organism and a list of annotated genes in the organism and output the enrichment p-value for each unique GO term in the input annotation. Enriched GO terms are color mapped in GO visualizer. User can also adjust the number of enriched GO terms to visualize.

In the Protein Set tab, users can input a list of annotated proteins and NaviGO will calculate the functional similarity between each pair of input proteins using Funsim score [8, 17] with different similarity schemes similar as in the GO Set tab. The confidence of similarity predictions is classified into five levels: very high, high, moderate, low, and the rest. It indicates the score is within top 1%, 5%, 10%, and 20% relative to the score distribution of all protein pairs of an arbitrary organism specified by the user. The upper section in the result page shows an interactive clustering view based on protein similarity score (Fig. 6). A user-defined cutoff value controls the connectivity of edges between nodes, and scoring schemes can be switched using the bar on the top right-hand corner of the network panel. The computed analysis results can also be download as a table in the CSV format.

Acknowledgments

This work was supported partly by the National Institutes of Health (R01GM097528), the National Science Foundation (IIS1319551, DBI1262189, IOS1127027).

References

1. Altschul SF, Madden TL, Schaffer AA, Zhang J, Zhang Z, Miller W, Lipman DJ (1997) Gapped BLAST and PSI-BLAST: a new generation of protein database search programs. Nucleic Acids Res 25(17):3389–3402
2. Pearson WR (1990) Rapid and sensitive sequence comparison with FASTP and FASTA. Methods Enzymol 183:63–98

3. Hawkins T, Kihara D (2007) Function prediction of uncharacterized proteins. J Bioinforma Comput Biol 5(1):1–30
4. Sael L, Chitale M, Kihara D (2012) Structure- and sequence-based function prediction for non-homologous proteins. J Struct Funct Genom 13(2):111–123. doi:10.1007/s10969-012-9126-6
5. Radivojac P, Clark WT, Oron TR, Schnoes AM, Wittkop T, Sokolov A, Graim K, Funk C, Verspoor K, Ben-Hur A, Pandey G, Yunes JM, Talwalkar AS, Repo S, Souza ML, Piovesan D, Casadio R, Wang Z, Cheng J, Fang H, Gough J, Koskinen P, Toronen P, Nokso-Koivisto J, Holm L, Cozzetto D, Buchan DWA, Bryson K, Jones DT, Limaye B, Inamdar H, Datta A, Manjari SK, Joshi R, Chitale M, Kihara D, Lisewski AM, Erdin S, Venner E, Lichtarge O, Rentzsch R, Yang H, Romero AE, Bhat P, Paccanaro A, Hamp T, Kaszner R, Seemayer S, Vicedo E, Schaefer C, Achten D, Auer F, Boehm A, Braun T, Hecht M, Heron M, Honigschmid P, Hopf TA, Kaufmann S, Kiening M, Krompass D, Landerer C, Mahlich Y, Roos M, Bjorne J, Salakoski T, Wong A, Shatkay H, Gatzmann F, Sommer I, Wass MN, Sternberg MJE, Skunca N, Supek F, Bosnjak M, Panov P, Dzeroski S, Smuc T, Kourmpetis YAI, van Dijk ADJ, Braak CJF, Zhou Y, Gong Q, Dong X, Tian W, Falda M, Fontana P, Lavezzo E, Di Camillo B, Toppo S, Lan L, Djuric N, Guo Y, Vucetic S, Bairoch A, Linial M, Babbitt PC, Brenner SE, Orengo C, Rost B, Mooney SD, Friedberg I (2013) A large-scale evaluation of computational protein function prediction. Nat Methods 10 (3):221–227. http://www.nature.com/nmeth/journal/v10/n3/abs/nmeth.2340.html supplementary-information
6. Jiang Y, Ronnen Oron T, Clark WT, Bankapur AR, D'Andrea D, Lepore R, Funk CS, Kahanda I, Verspoor KM, Ben-Hur A, Koo E, Penfold-Brown D, Shasha D, Youngs N, Bonneau R, Lin A, Sahraeian SM, Martelli PL, Profiti G, Casadio R, Cao R, Zhong Z, Cheng J, Altenhoff A, Skunca N, Dessimoz C, Dogan T, Hakala K, Kaewphan S, Mehryary F, Salakoski T, Ginter F, Fang H, Smithers B, Oates M, Gough J, Törönen P, Koskinen P, Holm L, Chen C-T, Hsu W-L, Bryson K, Cozzetto D, Minneci F, Jones DT, Chapman S, Dukka BKC, Khan IK, Kihara D, Ofer D, Rappoport N, Stern A, Cibrian-Uhalte E, Denny P, Foulger RE, Hieta R, Legge D, Lovering RC, Magrane M, Melidoni AN, Mutowo-Meullenet P, Pichler K, Shypitsyna A, Li B, Zakeri P, ElShal S, Tranchevent L-C, Das S, Dawson NL, Lee D, Lees JG, Sillitoe I, Bhat P, Nepusz T, Romero AE, Sasidharan R, Yang H, Paccanaro A, Gillis J, Sedeño-Cortés AE, Pavlidis P, Feng S, Cejuela JM, Goldberg T, Hamp T, Richter L, Salamov A, Gabaldon T, Marcet-Houben M, Supek F, Gong Q, Ning W, Zhou Y, Tian W, Falda M, Fontana P, Lavezzo E, Toppo S, Ferrari C, Giollo M, Piovesan D, Tosatto S, del Pozo A, Fernández JM, Maietta P, Valencia A, Tress ML, Benso A, Di Carlo S, Politano G, Savino A, Rehman HU, Re M, Mesiti M, Valentini G, Bargsten JW, van Dijk AD, Gemovic B, Glisic S, Perovic V, Veljkovic V, Veljkovic N, Almeida-e-Silva DC, Vencio RZ, Sharan M, Vogel J, Kansakar L, Zhang S, Vucetic S, Wang Z, Sternberg MJ, Wass MN, Huntley RP, Martin MJ, O'Donovan C, Robinson PN, Moreau Y, Tramontano A, Babbitt PC, Brenner SE, Linial M, Orengo CA, Rost B, Greene CS, Mooney SD, Friedberg I, Radivojac P (2016) An expanded evaluation of protein function prediction methods shows an improvement in accuracy. Genome Biol 17 (1):184. doi:10.1186/s13059-016-1037-6
7. Hawkins T, Luban S, Kihara D (2006) Enhanced automated function prediction using distantly related sequences and contextual association by PFP. Protein Sci 15 (6):1550–1556. doi:10.1110/ps.062153506
8. Hawkins T, Chitale M, Luban S, Kihara D (2009) PFP: automated prediction of Gene Ontology functional annotations with confidence scores using protein sequence data. Proteins 74(3):566–582. doi:10.1002/prot.22172
9. Chitale M, Hawkins T, Park C, Kihara D (2009) ESG: extended similarity group method for automated protein function prediction. Bioinformatics 25(14):1739–1745. doi:10.1093/bioinformatics/btp309
10. Seok YJ, Sondej M, Badawi P, Lewis MS, Briggs MC, Jaffe H, Peterkofsky A (1997) High affinity binding and allosteric regulation of *Escherichia coli* glycogen phosphorylase by the histidine phosphocarrier protein, HPr. J Biol Chem 272(42):26511–26521
11. D'Ari L, Rabinowitz JC (1991) Purification, characterization, cloning, and amino acid sequence of the bifunctional enzyme 5,10-methylenetetrahydrofolate dehydrogenase/5,10-methenyltetrahydrofolate cyclohydrolase from *Escherichia coli*. J Biol Chem 266 (35):23953–23958
12. Khan IK, Wei Q, Chapman S, Kc DB, Kihara D (2015) The PFP and ESG protein function prediction methods in 2014: effect of database updates and ensemble approaches. GigaScience 4:43. doi:10.1186/s13742-015-0083-4
13. Chitale M, Khan IK, Kihara D (2013) In-depth performance evaluation of PFP and ESG

sequence-based function prediction methods in CAFA 2011 experiment. BMC Bioinform 14(Suppl 3):S2. doi:10.1186/1471-2105-14-S3-S2

14. Lopez G, Rojas A, Tress M, Valencia A (2007) Assessment of predictions submitted for the CASP7 function prediction category. Proteins 69(Suppl 8):165–174. doi:10.1002/prot.21651
15. Khan IK, Wei Q, Chitale M, Kihara D (2015) PFP/ESG: automated protein function prediction servers enhanced with Gene Ontology visualization tool. Bioinformatics 31 (2):271–272. doi:10.1093/bioinformatics/btu646
16. Shannon P, Markiel A, Ozier O, Baliga NS, Wang JT, Ramage D, Amin N, Schwikowski B, Ideker T (2003) Cytoscape: a software environment for integrated models of biomolecular interaction networks. Genome Res 13 (11):2498–2504. doi:10.1101/gr.1239303
17. Schlicker A, Domingues FS, Rahnenfuhrer J, Lengauer T (2006) A new measure for functional similarity of gene products based on Gene Ontology. BMC Bioinform 7:302. doi:10.1186/1471-2105-7-302
18. Chitale M, Palakodety S, Kihara D (2011) Quantification of protein group coherence and pathway assignment using functional association. BMC Bioinform 12:373–373. doi:10.1186/1471-2105-12-373
19. Yerneni S, Khan I, Wei Q, Kihara D (2015) IAS: interaction specific GO term associations for predicting protein–protein interaction networks. IEEE/ACM Trans Comput Biol Bioinform. doi:10.1109/TCBB.2015.2476809
20. Sánchez J, Mardia KV, Kent JT, Bibby JM (1982) Multivariate analysis. Academic Press, London-New York-Toronto-Sydney-San Francisco 1979. xv, 518 pp., $ 61.00. Biom J 24 (5):502–502. doi:10.1002/bimj.4710240520

Chapter 2

GHOSTX: A Fast Sequence Homology Search Tool for Functional Annotation of Metagenomic Data

Shuji Suzuki, Takashi Ishida, Masahito Ohue, Masanori Kakuta, and Yutaka Akiyama

Abstract

Metagenomic analysis based on whole genome shotgun sequencing data requires fast protein sequence homology searches for predicting the function of proteins coded on metagenome short reads. However, huge amounts of sequence data cause even general homology search analyses using BLASTX to become difficult in terms of computational cost. GHOSTX is a sequence homology search tool specifically developed for functional annotation of metagenome sequences. The tool is more than 160 times faster than BLASTX and has sufficient search sensitivity for metagenomic analysis. Using this tool, user can perform functional annotation of metagenomic data within a short time and infer metabolic pathways within an environment.

Keywords Metagenomic analysis, Sequence homology search, Whole genome shotgun sequencing, Functional annotation, Substitution-score matrix

1 Introduction

Metagenomics is the study of the genomes of uncultured microbes obtained directly from microbial communities in their natural habitats. Such analyses have recently become more popular and important as the throughput of DNA sequencers has increased. Previously, metagenomic analysis was performed based on 16S rRNA data obtained from Sanger-sequencing methods, and the aim was to obtain the phylogenetic profiles of microbial communities from a target environment. However, whole-genome shotgun (WGS) sequencing, carried out using next-generation sequencing (NGS) technologies, produces huge amounts of metagenomic data. This enables us to uncover an abundance of orthologous groups, i.e., the distribution of gene/protein functions, in environmental samples. Based on such information, we can infer metabolic pathways within an environment and compare a metagenomic sample to the others based on its functions or functional

Daisuke Kihara (ed.), *Protein Function Prediction: Methods and Protocols*, Methods in Molecular Biology, vol. 1611, DOI 10.1007/978-1-4939-7015-5_2, © Springer Science+Business Media LLC 2017

categories. Various metagenomic studies have employed functional annotation techniques, facilitating novel scientific discoveries, such as the existence of enterotypes [1] and the relationship between gut microbes and type II diabetes [2].

To perform functional annotations of metagenomic data, it is necessary to determine the function of DNA short reads obtained from environmental samples. However, such metagenome sequences generally include DNA sequences from many different species, and closely related reference genome sequences are often unavailable. Thus, DNA short reads are translated into protein-coding sequences, and the functions of proteins are then predicted for more sensitive identification of novel genes. The BLASTX [3] program and well-annotated protein sequence databases, such as KEGG [4] and COG [5], have been used for protein function prediction with metagenomic data [6]. However, the computing speed of BLASTX is insufficient for analyzing the large quantities of metagenomic data produced by current sequencers, such as the Illumina HiSeq2500, which can produce several hundred billion base pairs of sequence data in a single run. Thus, a special tool is required for functional prediction of metagenome sequences.

GHOSTX is a sequence homology search tool specifically developed for functional annotation of metagenome sequences [7]. The GHOSTX algorithm uses a seed search method that relies on a score-based optimal seed length. In this method, only subsequences with scores greater than or equal to a threshold are searched, based on a given score matrix. Thus, the algorithm can effectively exclude seeds with sufficient length but insufficient match scores. In addition, the program accelerates its seed search process by using suffix arrays of both queries and database sequences. As a result, GHOSTX can achieve approximately 160 times greater speeds than BLASTX searches at similar levels of sensitivity. Several other tools, such as RAPSearch2 [8] and DIAMOND [9], have also been developed for functional annotation of metagenome sequences. The performances of these programs are comparable to that of GHOSTX; however, their algorithms are designed on the premise of the BLOSUM62 score matrix, whereas GHOSTX can use any score matrix and thus can be applied in various types of analyses. The GHOSTX program has been employed for several services, such as GhostKOALA [10], and has been shown to provide reliable results.

2 Materials

In this section, we describe the input data to execute GHOSTX and supplemental tools for the analysis of GHOSTX output data.

2.1 Input Data

The input data for GHOSTX are metagenome sequences produced from a DNA sequencer as query and protein sequences as a

```
>61JGYAAXX100510:7:115:4328:19392/1
ACGTTCCCCTAAGTCCAATGCTTTGAACGCTTGGCCATATTGCTCAAGCACTTGTGATTC
TTTGATTGTGTAATCACCG
>61JGYAAXX100510:7:116:12157:10767/1
TGTCAGTTCCTTCAAAGGTATCCAAGTCAAAGGCATGCATTGGTTGACCAAAGTAGAGCA
GGATGTAGTTGGTCACGTCCACTACGTTATTGATGGGACGA
>61JGYAAXX100510:7:116:19272:8295/1
ATCTCCAAAGCAGCAAGTGTCGCAAGTGATATCCACGCTTCAGGATCAGTAATCCATTCA
AACATAATTTTTCCTATAATAGAGTTAAATAAATCCGAAAG
>61JGYAAXX100510:7:116:9375:5125/1
CAACAATACTATGTCATTGTGAAAGCGAAAGAACTTCCAGTTGTACCAACAAAACCAGTA
AATAAAAATAATGAACCAGTAGATCCAACAGATCCAACAAC
```

Fig. 1 An example of a FASTA-formatted file. The description line is distinguished from the sequence data by a greater-than (">") symbol at the beginning of a line. By concatenating multiple single-sequence FASTA files, a multiple sequence FASTA formatted-file can be generated. The figure is a part of deep WGS sequencing data from the Human Microbiome Project (HMP) [11], with sequences from a buccal mucosa sample (SRS011090)

database. As a query input, GHOSTX requires the DNA sequence data in a FASTA format (Fig. 1). A multiple sequence FASTA format is acceptable. If the format of a query file is FASTQ, the user has to use an external tool, such as FASTX-Toolkit [12], to convert the format. The `fastq_to_fasta` command of FASTX-Toolkit converts the FASTQ-formatted file into a FASTA-formatted file.

```
$ fastq_to_fasta -i query.fastq -o query.fasta
```

The database sequences of GHOSTX are annotated protein sequences in the FASTA format. The database must be indexed preliminarily. The steps required to index the database sequences are described in Subheading 3. The user can use any protein sequence database, such as NCBI nr or Uniprot. However, well-annotated databases, such as COG and EggNOG, are recommended for functional prediction and analysis. In addition, the KEGG GENES database [4] is preferable because the post-analysis tool KEGG Analyzer, developed by our laboratory, is compatible only with the KEGG database.

2.2 Programs in the Software Packages

GHOSTX is available as open-source free software under the terms of the BSD 2-Clause license in source code form. GHOSTX is written in C++. It can be compiled and run on a wide variety of UNIX platforms and similar systems (including FreeBSD and Linux). The core of the GHOSTX is `ghostx` binary, generated after source codes are compiled. The KEGG Analyzer program for phylogenetic analysis and functional analysis is also available as open-source free software.

2.3 Web Sites

GHOSTX (version 1.3.7) is available at http://www.bi.cs.titech.ac.jp/ghostx/, and the previous version is still downloadable at http://www.bi.cs.titech.ac.jp/ghostx/releases/. The KEGG analyzer program is also available at http://www.bi.cs.titech.ac.jp/ghostx/kegg/.

3 Methods

3.1 Database Construction

Similar to NCBI BLAST, database sequences have to be indexed before GHOSTX sequence homology search. To construct the indices of database sequences, the `ghostx db` command is used.

```
$ ghostx db -i database.fasta -o exdb
```

The `ghostx db` command converts a FASTA file into GHOSTX format database files. The input format of `ghostx db` is as follows:

```
$ ghostx db -i database_fasta_file -o database_
name [-l chunk_size] [-t p|d]
```

where `-i` and `-o` options are required. `-l` specifies the chunk size of a database (default as 1 GByte = "`-l 1073741824`"). GHOSTX requires a considerable amount of memory during its homology search. The required memory size depends on the database size. And the indexing technique requires larger memory size than storing original sequences. However, current computing systems generally have relatively small memory sizes compared with database sizes. Therefore, GHOSTX divides a database into several chunks, each of whose size is specified on the `-l` option, before it constructs its indices. For example, when using the KEGG GENES database (3.9 GByte), with the chunk sizes set to 512 Mbyte, 1 GByte, and 2 GByte, the total memory sizes required for storing indexed database and performing homology searches are 4.6, 9.2, and 18.2 GByte, respectively. The relative computation speeds (based on a 2 GByte chunk) are 0.8×, 0.9×, and 1.0× when the chunk sizes are set to 512 Mbyte, 1 GByte, and 2 GByte, respectively [7]. The `-t` option can designate the database sequence type as protein "`p`" or DNA "`d`." Protein "`p`" is chosen as default. Searching the query sequence on a DNA database can be executed by "`-t d`."

3.2 Homology Search

To perform a sequence homology search, the `ghostx aln` command is used.

```
$ ghostx aln -i query.fasta -d exdb -o exout
```

The `ghostx aln` command has several options; the required arguments are `-i` as input query fasta, `-d` as indexed database file, and `-o` as output file name, and the additional options are as follows:

`-a`	number of CPU threads (default: 1)
`-v`	maximum number of alignments for each subject
`-b`	maximum number of the output for a query
`-q`	query sequence type, protein "`p`" or DNA "`d`" (default: "`d`")
`-t`	database sequence type, protein "`p`" or DNA "`d`" (default: "`p`")
`-s`	upper mismatch score in seed extension
`-T`	threshold score of seed search
`-G`	gap opening penalty (same as BLAST)
`-E`	gap extension penalty (same as BLAST)
`-F`	filter query sequence (same as BLAST)
`-M`	score matrix file

Among these parameters, the upper mismatch score D (`-s`) is the limit of acceptable score difference to determine whether a seed extends or not, and the threshold of seed search T_{seed} (`-T`) is minimum score for a hit in seed search. They regulate the sensitivity and computation speed of searches. The default parameters of "`-s 4 -T 30`" ($D = 4$, $T_{seed} = 30$) have the best balance between the sensitivity and computation speed. At this time, the sensitivity of GHOSTX is almost the same as that of RAPSearch2 [8]. If a faster calculation is needed, parameters of "`-s 1, -T 30`" ($D = 1$, $T_{seed} = 30$) with a smaller mismatch allowance are a good option. On the other hand, if higher sensitivity is needed, "`-s 4, -T 24`" ($D = 4$, $T_{seed} = 24$) with a lower threshold can be used. Refer to Table S1 of [7] for additional information regarding the relationships of these cutoff parameters with the balance between sensitivity and calculation time.

Furthermore, the `-b` option can be used to specify the maximum number of outputs for a query, and the default is 10. The `-v` option can be used to specify the maximum number of alignments for each subject, and the default is 1. The `-F` option is for masking off segments of the query sequence that have low compositional complexity determined by the SEG program [13]. Note that the threshold *E*-value for saving hits (`-e` of legacy BLAST and `-evalue` of BLAST+) does not exist in GHOSTX; thus, if the user wants to eliminate hits that have higher *E*-values than a user-defined threshold some text processing is needed after obtaining the GHOSTX output.

1	2	3	4	5	6	7	8	9	10	11	12
61JGYAAXX100510:7:100:10000:7489/1	smb:smi_0286 hypothetical protein; K07024	0.969697	33	1	0	100	2	79	111	4.09365e-21	102.436
61JGYAAXX100510:7:100:10000:7489/1	std:SPPN_10260 putative HAD superfamily hydrolase; K07024	0.969697	33	1	0	100	2	79	111	5.49279e-21	102.012
61JGYAAXX100510:7:100:10000:7489/1	snx:SPNOXC_17560 haloacid dehalogenase-like hydrolase; K07024	0.939394	33	2	0	100	2	79	111	1.03901e-19	97.7704
61JGYAAXX100510:7:100:10000:7489/1	spv:SPH_2139 Cof family protein; K07024	0.939394	33	2	0	100	2	66	98	1.03901e-19	97.7704
61JGYAAXX100510:7:100:10000:7489/1	snm:SP70585_2069 Cof family protein; K07024	0.939394	33	2	0	100	2	79	111	1.03901e-19	97.7704
61JGYAAXX100510:7:100:10001:16821/1	snc:HMPREF0837_11837 hypothetical protein	0.888889	18	2	0	47	100	1	18	1.04419e-07	57.9001
61JGYAAXX100510:7:100:10001:16821/1	spn:SP_1601 hypothetical protein	0.888889	18	2	0	47	100	1	18	1.04419e-07	57.9001
61JGYAAXX100510:7:100:10001:16821/1	snb:SP670_1685 hypothetical protein	0.888889	18	2	0	47	100	1	18	1.04419e-07	57.9001
61JGYAAXX100510:7:100:10001:16821/1	smb:smi_0674 hypothetical protein	0.888889	18	2	0	47	100	1	18	1.04419e-07	57.9001
61JGYAAXX100510:7:100:10001:16821/1	spd:SPD_1426 hypothetical protein	0.888889	18	2	0	47	100	1	18	1.04419e-07	57.9001

Fig. 2 An example of a GHOSTX output file. The GHOSTX output is BLAST-like tab-separated format. This example is a search result of a buccal mucosa metagenome sample from the HMP (SRS011090) using the KEGG GENES database. The columns are as follows: (*1*) name of the query sequence; (*2*) name of the homolog sequence (subject); (*3*) sequence identity; (*4*) alignment length; (*5*) the number of mismatches in the alignment; (*6*) the number of gap openings in the alignment; (*7*) start position of the query in the alignment; (*8*) end position of the query in the alignment; (*9*) start position of the subject in the alignment; (*10*) end position of the subject in the alignment; (*11*) *E*-value; and (*12*) normalized score

Figure 2 shows an example of the output from a GHOSTX homology search. The output format of GHOSTX is a tab-separated format, similar to that of BLAST. The format contains 12 columns, as described in the legend of Fig. 2.

3.3 Post-Analysis (KEGG Analyzer)

The KEGG Analyzer can calculate the corrected relative abundance of molecular-level functions based on KEGG Orthology (KO) from the GHOSTX search output with the KEGG GENES database. The KEGG Analyzer can also be used for generating phylogenetic profiles. The tool is effective and easy to use but requires a KEGG subscription because it refers to KEGG-licensed files. For more information, refer to the help file in the KEGG Analyzer program.

The outputs of the KEGG Analyzer include a normalized KO count (`ko.csv`) and normalized phylogenetic profile (`phylogeny.csv`). The statistical analysis of KO-based molecular functions and phylogenetic analysis can be performed using these files. An example of functional analysis with the tool is given in Subheading 4.

4 Case Study

In this section, we show an example of a homology search with GHOSTX and phylogenetic profile analysis and functional analysis from the homology search results. This case study aimed to identify differences in the human oral bacterial flora from WGS metagenomic samples based on phylogeny and gene function. The input data were deep WGS sequencing data from an HMP buccal mucosa sample (SRS011090), including a total of 1,787,927 paired-end reads (file size of FASTQ: 449 MByte). The database used here was `genes.pep` from the KEGG GENES database (file size: approximately 5.8 GByte on February 2, 2016).

4.1 Homology Search with GHOSTX

To perform phylogenetic analysis and functional analysis of metagenomic data, sequence homology searches are required. The user has to create indices of a database using the `ghostx db` command, as shown in the previous section.

```
$ ghostx db -i genes.pep -o kegg.db
```

Then, the user executes the homology search with the `ghostx aln` command using indexed database `kegg.db`.

```
$ ghostx aln -i SRS011090.fasta -d kegg.db
-o SRS011090_out.csv
```

Even using GHOSTX, the search process requires more than 100 h with a workstation. Thus, a cluster system with multiple computing nodes is recommended to execute the process [14]. Finally, the user can obtain the homology search result `SRS011090_out.csv` in a tab-separated format, as shown in Fig. 2.

4.2 Phylogenetic Profile Analysis

Here, we show an example of calculation for relative abundance of phylogeny at the genus and phylum levels from the KEGG Analyzer output. The KEGG Analyzer was executed using the GHOSTX output file and some files from the KEGG database.

```
$ kegg_analyzer genes_list gi_taxid root_map
ko_enzyme uscg_list SRS011090_out.csv
[ghostx_score_threshold] [ghostx_identity_
threshold] ko.csv uscg_count.csv normalize_
error.txt unnormalized_out.csv phylogeny.csv
```

where `genes_list` and `root_map` are generated by the scripts from KEGG and NCBI files contained in the KEGG Analyzer package, and `gi_taxid` and `ko_enzyme` are downloaded from KEGG FTP. `uscg_list` is contained in the KEGG Analyzer package. `ko.csv`, `uscg_count.csv`, `normalize_error.txt`, `unnormalized_out.csv`, and `phylogeny.csv` are output files. Only `phylogeny.csv` and `ko.csv` are used in this case study.

The user can generate a two-column file from `phylogeny.csv` using a one-line command as shown below:

Relative abundance based on genus rank:

```
$ tail -n +7 phylogeny.csv | cut -f 4,5 | awk -F
"\t" '{sum[$1]+=$2}END{for(k in sum){print k
"\t" sum[k]}}'
```

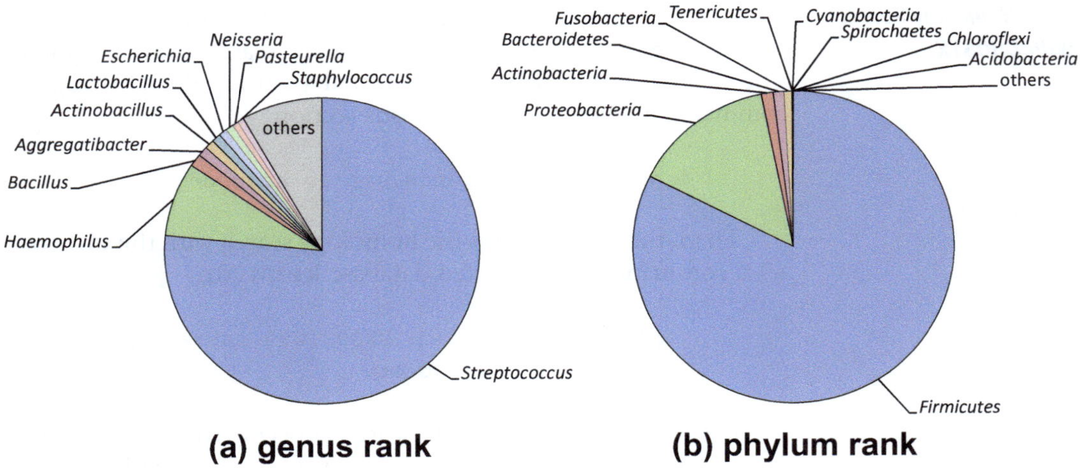

Fig. 3 Phylogenetic profiles of the buccal mucosa HMP metagenome sample (SRS011090). These pie charts show the top 10 phylogenies with higher abundance. The remaining phylogenies are grouped into "others"

Relative abundance based on phylum rank:

```
$ tail -n +7 phylogeny.csv | cut -f 3,5 | awk -F
"\t" '{sum[$1]+=$2}END{for(k in sum){print k
"\t" sum[k]}}'
```

Figure 3 describes the relative abundances as pie charts created using spreadsheet software. From the figure, we can understand that *Firmicutes* is the most major in the phylum rank and most of genera belonging to *Firmicutes* are *Streptococcus* in buccal mucosa.

4.3 Functional Analysis

The abundance and distribution of molecular functions (KOs) of metagenomic data can be understood from a normalized KO count (`ko.csv`) generated by KEGG Analyzer.

However, it is difficult to understand which pathway is specifically activated or inactivated only from the counts. In such a situation, mapping the information onto pathway maps improves our understanding.

iPATH2 [15] is a web-based tool for the visualization, analysis, and customization of various pathways maps. To assign the information onto the pathway map, the user should first access the iPATH2 website (http://pathways.embl.de/iPath2.cgi) and click the "Customize" button. Next, the user should click the "New selection" tab and paste the KO list in "Element selection" box. Then, the user should click the "Submit data and customize maps" button (Fig. 4). Parameters such as line color (# and hex color code), line width (W and integer value), and others can be set using the KO list.

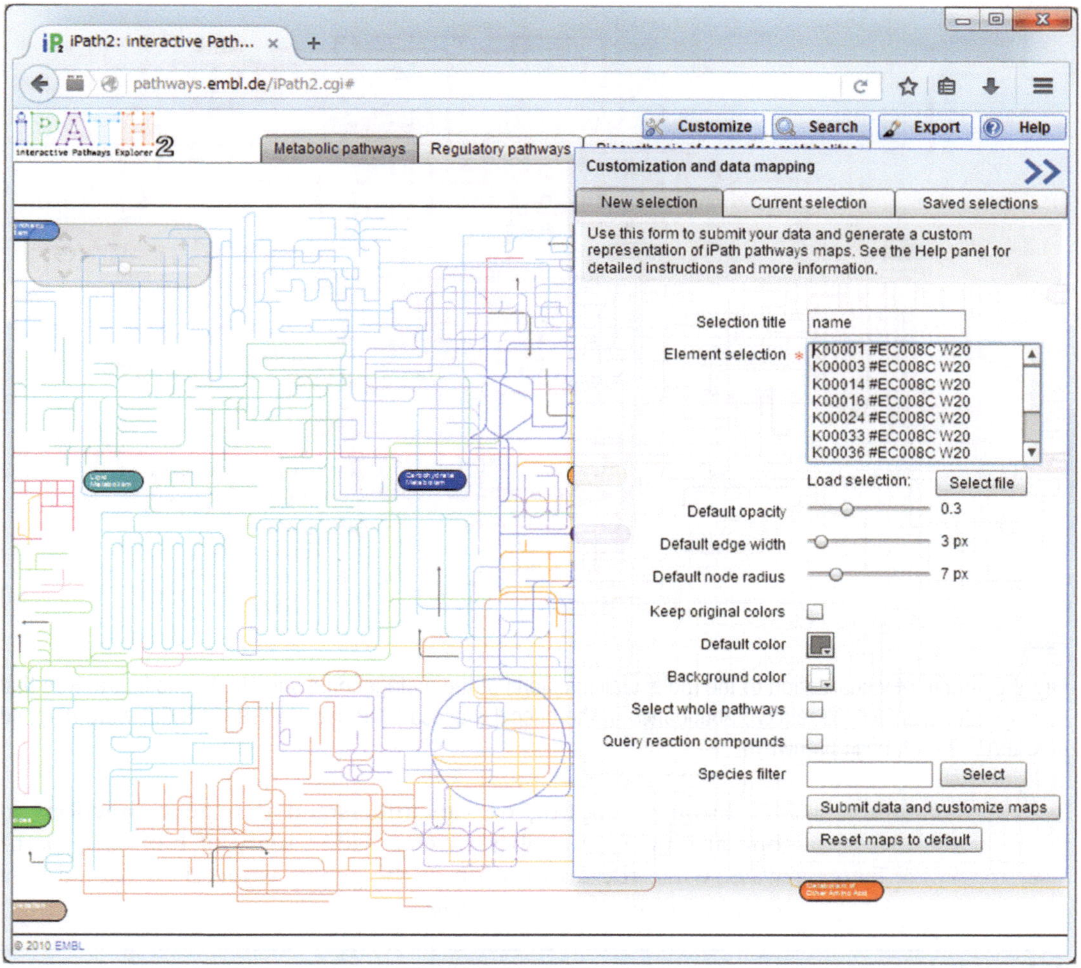

Fig. 4 Screen shot of iPATH2 element selection input. The KO list (color code of magenta [#EC008C] and line width [W20] are shown) is input into the "Element selection" box or "Load selection" dialogue as a text file

Figure 5 shows the KOs with greater than 0.01% relative abundance in the buccal mucosa HMP sample (SRS011090) loaded onto the iPATH2 pathway map. The iPATH2 input file was generated from the KEGG Analyzer output file (`ko.csv`) by a one-line command:

```
$ tail -n +3 ko.csv | awk -F "\t" '$4>0.0001 &&
$6!="KO not found"{print substr($6,4) " #EC008C
W20"}' > ipath.in
```

where `ipath.in` is the input file for iPATH2 mapping.

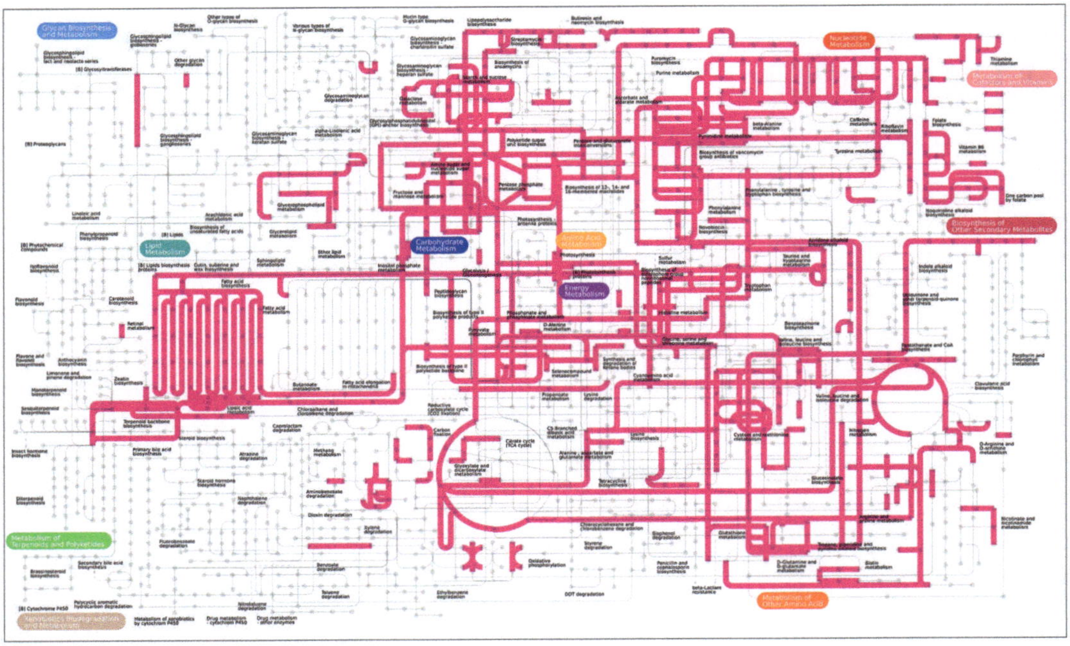

Fig. 5 Example of visualization of the molecular functions of the metagenome sample. In this figure, KOs that have greater than 0.01% relative abundance in the buccal mucosa HMP sample (SRS011090) were mapped onto iPATH2 metabolic pathways

From the figure, we can understand that gene functions in buccal mucosa microbes cover a wide variety of biological pathways, while the distribution of the genera is mostly occupied by a dominant genus, *Streptococcus.*

Acknowledgments

This work was partly supported by the Strategic Programs for Innovative Research (SPIRE) Field 1 Supercomputational Life Science of the Ministry of Education, Culture, Sports, Science and Technology (MEXT) of Japan and Core Research for Evolutional Science and Technology (CREST) "Extreme Big Data" of the Japan Science and Technology Agency (JST).

References

1. Arumugam M, Raes J, Pelletier E et al (2011) Enterotypes of the human gut microbiome. Nature 473:174–180
2. Qin J, Li Y, Cai Z et al (2012) A metagenome-wide association study of gut microbiota in type 2 diabetes. Nature 490:55–60
3. Altschul SF, Gish W, Miller W, Myers EW, Lipman DJ (1990) Basic local alignment search tool. J Mol Biol 215:403–410
4. Kanehisa M, Goto S (2000) KEGG: kyoto encyclopedia of genes and genomes. Nucleic Acids Res 28:27–30

5. Tatusov RL, Fedorova ND, Jackson JD et al (2003) The COG database: an updated version includes eukaryotes. BMC Bioinformatics 4:41
6. Kurokawa K, Itoh T, Kuwahara T et al (2007) Comparative metagenomics revealed commonly enriched gene sets in human gut microbiomes. DNA Res 14:169–181
7. Suzuki S, Kakuta M, Ishida T, Akiyama Y (2014) GHOSTX: an improved sequence homology search algorithm using a query suffix array and a database suffix array. PLoS ONE 9: e103833
8. Zhao Y, Tang H, Ye Y (2012) RAPSearch2: a fast and memory-efficient protein similarity search tool for next-generation sequencing data. Bioinformatics 28:125–126
9. Buchfink B, Xie C, Huson DH (2015) Fast and sensitive protein alignment using DIAMOND. Nat Methods 12:59–60
10. Kanehisa M, Sato Y, Morishima K (2016) BlastKOALA and GhostKOALA: KEGG tools for functional characterization of genome and metagenome sequences. J Mol Biol 428:726–731
11. The Human Microbiome Project Consortium (2012) Structure, function and diversity of the healthy human microbiome. Nature 486:207–214
12. Gordon A, Hannon G (2012) FASTX-Toolkit. http://hannonlab.cshl.edu/fastx_toolkit. Accessed 27 Mar 2016
13. Wootton JC, Federhen S (1993) Statistics of local complexity in amino acid sequences and sequence databases. Comput Chem 17:149–163
14. Kakuta M, Suzuki S, Ishida T, Akiyama Y. A massively parallel sequence similarity search for metagenomic sequencing data (submitted for publication)
15. Yamada T, Letunic I, Okuda S, Kanehisa M, Bork P (2011) iPath2.0: interactive pathway explorer. Nucleic Acids Res 39:W412–W415

Chapter 3

From Gene Annotation to Function Prediction for Metagenomics

Fatemeh Sharifi and Yuzhen Ye

Abstract

Microbes play important roles in almost every aspect of life, including human health and diseases. Facilitated by the rapid development of sequencing technologies, metagenomics research has accelerated the accumulation of genomic sequences of microbial species that had been inaccessible before. Analysis of the metagenomic sequencing data can reveal not only the species but also the functional composition of microbial communities. Here, we report a pipeline for functional annotation of metagenomic datasets. The pipeline is built from several programs that we have developed for metagenomic sequence analysis including a protein-coding gene predictor for short reads (or contigs) and a fast similarity search tool. Given a metagenomic dataset, the pipeline reports putative protein-coding genes (or gene fragments) and functional annotations of the genes in Gene Ontology (GO) terms and Enzyme Commission (EC) numbers, and potential metabolic pathways that are likely encoded by the metagenome. Fun4Me is available for download at https://sourceforge.net/projects/fun4me.

Keywords Metagenomics, Similarity search, Function prediction, Gene Ontology (GO), Metabolic pathway

1 Introduction

Due to the large volume of metagenomic datasets and their complexity (containing sequences from a mixture of species), functional annotation remains a challenging problem [1]. Over the years, we have developed several computational tools for metagenomic sequence analysis. Here, we report a pipeline, called Fun4Me, for functional annotation of metagenomes. Our pipeline takes a query metagenomic dataset (of either short sequencing reads, or assemblies) as the input, and outputs putative protein-coding genes or gene fragments, their putative functions in Gene Ontology (GO) terms [2] or EC numbers, and possible metabolic pathways.

The pipeline is built upon several computational tools that we have developed for metagenomic sequence analyses (*see* Fig. 1). The first tool is a predictor for protein coding genes,

Daisuke Kihara (ed.), *Protein Function Prediction: Methods and Protocols*, Methods in Molecular Biology, vol. 1611, DOI 10.1007/978-1-4939-7015-5_3,

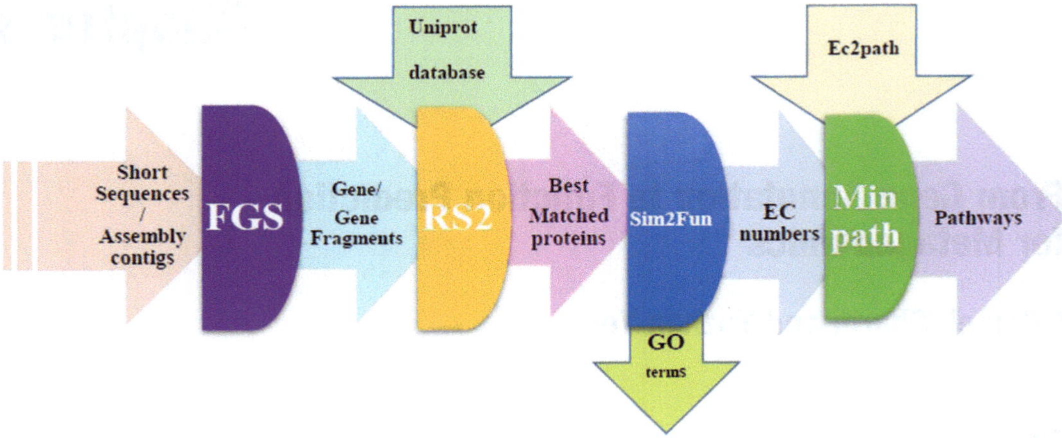

Fig. 1 Fun4Me workflow. Abbreviations: FGS (FragGeneScan); RS2 (RapSearch2)

FragGeneScan, which was developed to address the two challenging problems in gene prediction for metagenomic sequences: metagenomic sequences are short and error-prone, and are from many species [3]. FragGeneScan's core is a hidden Markov model (HMM), which incorporates codon usage bias, sequencing error models, and start/stop codon patterns in a unified model. FragGeneScan allows transitions between the insertion/deletion (indel) states and the match states, so it can effectively detect frameshifts that are caused by indel errors in sequencing. It predicts complete genes as well as partial (fragmented) genes without start and/or stop codons. The second tool RAPSearch2 achieves fast similarity search against reference protein database for metagenomic sequences, using reduced amino acid alphabet and flexible seed so that seeds of various lengths with mismatches can be identified quickly by hashing [4, 5]. Metagenomic datasets are getting bigger and bigger, and the homology search of the large metagenomic datasets against some reference database, often required by downstream analyses, has become a bottleneck in the analyses of metagenomic datasets. RAPSearch2 belongs to the new generation of computational tools for similarity searches [6] that are significantly faster than BLAST [7]. We note that a recently developed tool DIAMOND [8] achieves even faster search than RAPSearch2, but DIAMOND consumes large memory as compared to RAPSearch2 so is limited in this aspect. The third tool in the pipeline, MinPath, implements a parsimony approach to biological pathway reconstruction/inference for metagenomes [9]. Pathway analysis of metagenomic data involves characterization of the aggregate metabolic processes of microbial communities in a given environment. The incompleteness of the data makes it difficult to reconstruct the entire pathways encoded by a metagenome. We showed that MinPath achieves a more conservative, yet more faithful, estimation of the biological pathways for a query metagenomic dataset, and therefore the functionality of the corresponding microbial community [9].

Fun4Me pipeline can be used to infer protein-coding genes, and their putative functions from metagenomic datasets. These annotations can be used for pathway reconstruction and functional profiling of metagenomes, providing insights into the functionality of the corresponding microbial communities. The users can use our pipeline as a one-stop application, or use the individual tools in the pipeline for different purposes. We note all the tools (FragGeneScan, RAPSearch2, and MinPath) included in Fun4Me have been used by other researchers, either as individual tools, or as a tool embedded in their analysis workflows. For example, MinPath is used to identity minimum pathways in HUMAnN2 (The HMP Unified Metabolic Analysis Network 2) (http://huttenhower.sph.harvard.edu/humann2); FragGeneScan is used in MG-RAST (Metagenomics RAST server; http://metagenomics.anl.gov/) [10] as the gene caller; and RAPSearch2 is used as one of the similarity search engines (the other one is DIAMOND) in a recently developed tool SUPER-FOCUS for the fast functional analysis of shotgun metagenomic data [11].

2 Materials

Our pipeline for similarity-search-based functional prediction for metagenomic sequences includes several computational tools that we have developed. We include a similarity search database and accessory files needed, along with the pipeline, in a package called Fun4Me.

2.1 Programs Included in the Pipeline

1. *FragGeneScan*: a protein-coding gene predictor in short sequencing reads or contigs assembled from the sequencing reads. FragGeneScan is mainly implemented in C.
2. *RAPSearch2*: a tool for fast similarity search against a reference protein database. RAPSearch2 is implemented in C++.
3. *MinPath*: a tool providing conservative estimation of metabolic pathways based on the parsimony principle. MinPath tries to find the minimum pathways that can explain all the functions assigned to at least one protein predicted from the query dataset, which is formulated as an integer-programming problem [12]. It uses the GLPK package (GNU Linear Programming Kit; http://www.gnu.org/software/glpk/glpk.html) for solving the integer-programming problem; all the other functions are implemented in Python.

2.2 Similarity Search Database and Accessory Files Needed for Functional Annotation

1. *Similarity search database*: uniref-90-noE.fasta. It is the similarity search database that will be used by RAPSearch2. We downloaded a 90% nonredundant set of UniProt sequences (uniref90) from the Uniprot ftp website (ftp://ftp.uniprot.org). Since the main application of the similarity search is functional annotation, we only need to keep UniProt sequences that have GO assignments for similarity search. Also we excluded proteins of eukaryotic origin, considering that our pipeline will be mainly applied to annotate metagenomic datasets that only contain sequences from prokaryotes and/or viruses. By doing this, we significantly reduced the size of the search database, which is important for speeding up the similarity search. The resultant database (uniref90-go-noE.fasta) contains about 4.7 million proteins.
2. *Gene annotation file*: gene_association.goa_ref_uniprot. This file contains GO annotation of the UniProt proteins. It was downloaded from http://www.ebi.ac.uk/GOA. Similarly, we prepared a file gene_association.goa_ref_uniprot.noE containing GO associations for non-eukaryotic proteins.
3. *EC to GO mapping file*: ec2go. This file provides mapping between GO terms and EC numbers. It is used for EC assignments based on GO annotations. The file was downloaded from http://www.geneontology.org/external2go/ec2go.
4. *EC to pathway mapping file*: ec2path. This file was created using the files from the MetaCyc database (http://metacyc.org/download.shtml), pathways.dat and reactions.dat. Reactions inside each pathway were extracted and annotated with EC numbers.

2.3 Availability of the Package

The Fun4Me package, including source codes (implemented in C/C++ and Python) and data files mentioned above, is available for download at the Sourceforge website (https://sourceforge.net/projects/fun4me).

3 Methods

The users can download the Fun4Me package from its Sourceforge website and install it on a local Linux/Unix machine (*see* Subheading 3, **step 1**). The users can then call a wrapper script (fun4me.py) for one-step application of the package for functional annotation (*see* Subheading 3, **step 3**). However, the users may also follow individual steps (as shown in Subheading 3, **step 4**) so that different parameters or search databases can be used for their own purposes.

1. Installation. The users can call a script to install all the tools included in the pipeline. Once the package is downloaded, go under the root directory (Fun4Me), and call "./install."

2. Preprocessing of the similarity search database (uniref90-go-noE.rap) using "prerapsearch." Go to the data subfolder, and run the command "../tools/RAPSearch2.23_64bits/bin/prerapsearch -d uniref90-go-noE.fasta -n uniref90-go-noE.rap."
3. One-step application of the package for functional annotation using a wrapper script (fun4me.py). Given an input metagenomic dataset (*see* **Note 1**), the script invokes the multiple steps (*see* Subheading 3, **step 4**) for annotation, and produces outputs including putative protein coding genes (or gene fragments), similarity search results, GO and EC assignments, and metabolic pathways. For example, go to the tests subfolder, and run the command "../fun4me.py –i small.fa –o small," which takes small.fa as the input and produces ten output files, including small-fgs.faa (predicted proteins/protein fragments), small-fgs.gff (gene prediction results in the gff format), small-rap.m8 (the similarity search results), small-rap.go (the GO assignments), small-rap.ec (the EC assignments), small-rap.ec.minpath (the MinPath result), and small-pwy.html (a html report of the metabolic pathways).
4. Following individual steps.
   (a) *Gene prediction*: FragGeneScan will be called in this step. The input to FragGeneScan is a file of short sequences, or assembly contigs, in FASTA format (*see* **Note 2**). This step produces predicted protein-coding genes (or gene fragments; *see* **Note 3**) and their protein translation.
   (b) *Similarity search for the predicted proteins*: RAPSearch2 takes the output file of predicted protein sequences from **step 1** as input (*see* **Note 4**), searches them against the UniProt database (included in the package) (*see* **Note 5**), and outputs significant hits in a text file, one per line (*see* **Note 6**).
   (c) *GO and EC assignments based on similarity search results*: Two awk commands (implemented in fun4me.py) are used to assign GO terms and EC numbers to predicted genes, based on the similarity search result from **step 2** (*see* **Note 7**).
   (d) *Metabolic pathway reconstruction based on EC assignments*: MinPath takes the EC assignments as the input and identifies the list of pathways that are needed to explain all the annotated functions (*see* **Note 8**).
5. Case study. Here, we use a small dataset to demonstrate the utility of Fun4Me. This small dataset was prepared from a stool metagenomic dataset from the Human Microbiome Project (HMP) [13] with ID of SRS011061. We only used a small fraction of the reads for demonstration purposes (the reads file called small.fa can be found under the subfolder called tests in the package). Running Fun4Me on this small dataset results in ten files: most are pure text files (predicted genes, GO and EC

Metabolic pathway summary

Pathway	Description	#Functions	#Annotated	Functions
P101-PWY	ectoine biosynthesis	5	1	EC.1.2.1.11
P121-PWY	adenine and adenosine salvage I	2	1	EC.2.4.2.1
P184-PWY	protocatechuate degradation I (*meta*-cleavage pathway)	8	1	EC.4.1.1.3
P21-PWY	pentose phosphate pathway (partial)	2	1	EC.2.2.1.1
P224-PWY	sulfate reduction V (dissimilatory)	3	1	EC.2.7.7.4
P42-PWY	incomplete reductive TCA cycle	7	2	EC.1.2.7.3 EC.6.4.1.1
PEPTIDOGLYCANSYN-PWY	peptidoglycan biosynthesis I (*meso*-diaminopimelate containing)	2	2	EC.2.4.1.227 EC.2.7.8.13
PPGPPMET-PWY	ppGpp biosynthesis	4	1	EC.2.7.6.5
PROSYN-PWY	L-proline biosynthesis I	3	2	EC.1.2.1.41 EC.2.7.2.11

Fig. 2 A snapshot of the html report produced by Fun4Me for metabolic pathways

assignments, etc.) except one in html format, which summarizes the reconstructed metabolic pathways. The html file provides links to the pathways and enzymes, so interested users can follow the links to check the details of the pathways (and their diagrams) and enzymes. Figure 2 shows a snapshot of the html report of the pathways. The first pathway is ectoine biosynthesis (MetaCyc ID: P101-PWY, the first one in Fig. 2); the pathway involves five enzymes, but only one (EC.1.2.1.11) is annotated in the query dataset. Since the enzyme EC.1.2.1.11 can potentially be involved in other pathways (a total of seven MetaCyc pathways contain this enzyme), P101-PWY appears suspicious, even though MinPath selects it. The second pathway reported is adenine and adenosine salvage I (P121-PWY), which involves two enzymes. One of the enzymes, EC.2.4.2.1, has annotation in the query dataset by Fun4Me. Considering that EC.2.4.2.1 is unique to this pathway, and its annotation is supported by a significant similarity match (with an *E*-value of 8e-11), we are more confident with this pathway (P121-PWY) as compared to the first one (P101-PWY). We note that reconstructed pathways are very patchy for this case, because we used only a small fraction of the original metagenomic dataset as the query. However, pathways reconstructed for metagenomes will be unavoidably patchy (only some of the involved enzymes have annotations) as metagenomes are most likely to be incomplete.

4 Notes

1. *Input file:* Fun4Me is designed for annotation of metagenomic sequences, but it can also be used to annotate metatranscriptomes and other kinds of sequencing data. The users can also use -p option to specify protein sequences as the input.

2. *FragGeneScan parameters:* The advantage of using FragGeneScan is that users do not need to train specific models for gene prediction for their sequences. The wrapper script (fun4me.py) invokes FragGeneScan with parameters according to the -r parameter the users specify. If the users call FragGeneScan directly, they may choose to use "-w 1 -t complete" with FragGeneScan when they have the assembly of the metagenomic sequences as the input, to refine the prediction of start codes of putative protein-coding genes. Assemblers such as MEGAHIT [14] can be used to assemble metagenomes.
3. *Complete or fragmented genes*: Protein-coding genes found in short sequencing reads are most likely fragmented; and those found in contigs may also be fragmented. FragGeneScan predicts and outputs both complete and fragmented genes.
4. *Input to RAPSearch2*: The input can be protein sequences, or nucleotide sequences (which will be translated in six frames for protein similarity search).
5. *RAPSearch2 supports slow and fast search modes*: By default, RAPSearch2 will be called in slow mode. RAPSearch2 in fast mode uses several greedy strategies to avoid alignment extensions that are unlikely to result in meaningful alignment, and therefore speeds up the search significantly. The fast mode only causes minor loss of similarity search sensitivity, especially when the input sequences are short (e.g., the input sequences are short sequencing reads, or predicted protein fragments from short sequences).
6. *RAPSearch2 output*: By default, RAPSearch2 outputs the one-line descriptions of the best hits, similar to the BLAST output (−outfmt 6), except that it lists log(*E*-value) instead of *E*-value for each hit. The wrapper script fun4me.py calls RAPSearch2 using "-s T" option so that it outputs one-line descriptions using *E*-values as in BLAST.
7. *GO and EC assignments*: In current implementation of the pipeline, we use the top-hit approach for the functional assignments (i.e., for each query protein, the GO terms associated with the most similar subject sequence in the reference database will be transferred to the query protein). We will be incorporating algorithms such as those implemented in Argot [15] to consider all significant hits for GO assignments, which must be subject to a postprocessing to filter out less significant and less specific GO terms.
8. *Pathway collections*: We only provide analysis of metabolic pathways included in the MetaCyc database [16]. Our package does not include KEGG pathways (http://www.kegg.jp/kegg/). However if the users have access to KEGG pathways and associated KO families, they can choose to use KEGG pathways (MinPath works with any pathways, as long as the function to pathway mapping is available).

Acknowledgments

This work was supported by NIH grant 1R01AI108888 and NSF grant DBI-0845685.

References

1. Wooley JC, Ye Y (2009) Metagenomics: facts and artifacts, and computational challenges. J Comput Sci Technol 25(1):71–81
2. Gene Ontology C (2015) Gene Ontology Consortium: going forward. Nucleic Acids Res 43(Database issue):D1049–D1056
3. Rho M, Tang H, Ye Y (2010) FragGeneScan: predicting genes in short and error-prone reads. Nucleic Acids Res 38(20):e191
4. Ye Y, Choi JH, Tang H (2011) RAPSearch: a fast protein similarity search tool for short reads. BMC Bioinformatics 12:159
5. Zhao Y, Tang H, Ye Y (2012) RAPSearch2: a fast and memory-efficient protein similarity search tool for next-generation sequencing data. Bioinformatics 28(1):125–126
6. Somervuo P, Holm L (2015) SANSparallel: interactive homology search against Uniprot. Nucleic Acids Res 43(W1):W24–W29
7. Altschul SF, Madden TL, Schaffer AA, Zhang J, Zhang Z, Miller W, Lipman DJ (1997) Gapped BLAST and PSI-BLAST: a new generation of protein database search programs. Nucleic Acids Res 25(17):3389–3402
8. Buchfink B, Xie C, Huson DH (2015) Fast and sensitive protein alignment using DIAMOND. Nat Methods 12(1):59–60
9. Ye Y, Doak TG (2009) A parsimony approach to biological pathway reconstruction/inference for genomes and metagenomes. PLoS Comput Biol 5(8):e1000465
10. Meyer F, Paarmann D, D'Souza M, Olson R, Glass EM, Kubal M, Paczian T, Rodriguez A, Stevens R et al (2008) The metagenomics RAST server - a public resource for the automatic phylogenetic and functional analysis of metagenomes. BMC Bioinformatics 9:386
11. Silva GG, Green KT, Dutilh BE, Edwards RA (2016) SUPER-FOCUS: a tool for agile functional analysis of shotgun metagenomic data. Bioinformatics 32(3):354–361
12. Bertsimas D, Tsitsiklis JN (1997) Introduction to linear optimization. In: Athena Scientific series in optimization and neural computation. Athena Scientific, Belmont, MA. xv, 587 p
13. The Human Microbiome Project Consortium (2012) Structure, function and diversity of the healthy human microbiome. Nature 486 (7402):207–214
14. Li D, Luo R, Liu CM, Leung CM, Ting HF, Sadakane K, Yamashita H, Lam TW (2016) MEGAHIT v1.0: A fast and scalable metagenome assembler driven by advanced methodologies and community practices. Methods 102:3–11. doi:10.1016/j.ymeth.2016.02.020
15. Lavezzo E, Falda M, Fontana P, Bianco L, Toppo S (2016) Enhancing protein function prediction with taxonomic constraints—the Argot2.5 web server. Methods 93:15–23
16. Caspi R, Billington R, Ferrer L, Foerster H, Fulcher CA, Keseler IM, Kothari A, Krummenacker M, Latendresse M et al (2016) The MetaCyc database of metabolic pathways and enzymes and the BioCyc collection of pathway/genome databases. Nucleic Acids Res 44 (D1):D471–D480

Chapter 4

An Agile Functional Analysis of Metagenomic Data Using SUPER-FOCUS

Genivaldo Gueiros Z. Silva, Fabyano A.C. Lopes, and Robert A. Edwards

Abstract

One of the main goals in metagenomics is to identify the functional profile of a microbial community from unannotated shotgun sequencing reads. Functional annotation is important in biological research because it enables researchers to identify the abundance of functional genes of the organisms present in the sample, answering the question, "What can the organisms in the sample do?" Most currently available approaches do not scale with increasing data volumes, which is important because both the number and lengths of the reads provided by sequencing platforms keep increasing. Here, we present SUPER-FOCUS, SUbsystems Profile by databasE Reduction using FOCUS, an agile homology-based approach using a reduced reference database to report the subsystems present in metagenomic datasets and profile their abundances. SUPER-FOCUS was tested with real metagenomes, and the results show that it accurately predicts the subsystems present in the profiled microbial communities, is computationally efficient, and up to 1000 times faster than other tools. SUPER-FOCUS is freely available at http://edwards.sdsu.edu/SUPERFOCUS.

Keywords Bioinformatics, Metagenomics, Functional profiling, Agile tool, Sensitive, SEED

1 Introduction

In many environments, 99% of the lineages in the resident microbial community cannot be cultured [1]. Metagenomics is a powerful tool that allows the study of environmental communities without this culture-bias and understands the real diversity present in microorganisms just by using rapidly developing DNA sequencing approaches [1]. Next-generation sequencing (NGS) technologies have accelerated the DNA sequencing process, decreased the cost, and opened new horizons in the environmental understanding [2].

Understanding microbial communities is important in many areas of biology. For example, it discriminates taxonomic and functional profiles of microbes associated with human [3] and global ocean microbiomes [4].

Daisuke Kihara (ed.), *Protein Function Prediction: Methods and Protocols*, Methods in Molecular Biology, vol. 1611, DOI 10.1007/978-1-4939-7015-5_4, © Springer Science+Business Media LLC 2017

Functional profiling of metagenomic sequences normally requires the alignment of sequences to a curated large database of annotated sequences to identify similar matches [5]. There are many databases for functional annotations, including the SEED [6], which contains subsystems (protein families with a similar function), and the large metabolic pathway databases KEGG [7] and MetaCyc [8].

Most of the available tools use either homology, i.e., where the metagenomic sequencing reads are aligned against a curated reference database, or exact matches via *k*-mers to associate metagenomic reads to the database sequences. Homology-based approaches normally use BLAST [9] or BLAT [10] to find the best hit in a reference database; however, those algorithms are slow and were not designed with NGS data in consideration. New homology search methods such as RAPSearch2 [11] and DIAMOND [12] have recently been developed to reduce the analysis run time. MEGAN [13] and MG-RAST [14] align query sequences to a reference database to profile the metagenomic data. MEGAN inputs the tabular results files created from programs such as BLAST, DIAMOND, or RAPSearch2 to the nonredundant database, and creates taxonomic and/or functional profiles. MG-RAST predicts the open reading frames (ORFs) on the metagenomic data and aligns them to the M5 nonredundant database using BLAT.

K-mer-based approaches use short sequences to find exact matches in a metagenomic query. For instance, real-time metagenomics (RTMg) [15] identifies all words of length *k* (between 7 and 12 amino acids) for a set of functionally related proteins, and uses them to find exact matches in the query to identify the functions present in the metagenomic sample.

Here, we present SUPER-FOCUS [16], SUbsystems Profile by databasE Reduction using FOCUS, which classifies each sequence in the metagenome into a subsystem by aligning all the input data against a reduced database. The speedup compared to other available tools comes from three improvements compared to the standard metagenome annotation pipelines. First, a clustered version of the SEED database that reduces the total search space; second, identification of the genera present in the metagenomic sample using FOCUS [17]; and finally, alignment of input data against the reduced database using RAPSearch2 which is 100 times faster than blastx [18]. SUPER-FOCUS is up to 1000 times faster than other tools with little loss of sensitivity, and still computationally efficient.

2 Materials

2.1 Input Data

The input data of SUPER-FOCUS is a file with DNA shotgun sequences in FASTA or FASTQ format. The program also allows the user to use proteins sequences as an input query; however, some parameters should be changed (please read "Running SUPER-FOCUS" below).

2.2 Dependencies

SUPER-FOCUS requires a set of packages and programs to be installed to run the tool, which are described in Table 1.

2.3 Aligner Choice

SUPER-FOCUS provides three options for aligning sequences: RAPSearch2 [11], DIAMOND [12], and BLAST [9]. The aligner choice should be made based on the user's computing resources. For example, RAPSearch2 should be the choice for users who have less memory available (about 2 GB RAM) and DIAMOND for the users who have at least 100 GB RAM. BLAST is known to be slow; therefore, it should be the choice when sensitivity is a factor in the analysis rather than speed.

DIAMOND is about 20 times faster than RAPSearch2 for large datasets, but slower than blastx for small datasets. RAPSearch2 and DIAMOND have fast versions, which are less sensitive but are 900 and 20,000 times faster than blastx respectively. Figure 1 presents a decision flowchart that guides the user on how to choose the proper aligner based on the input and hardware available.

2.4 Database Choice

SUPER-FOCUS requires a group of reference sequences to identify the subsystems present in a metagenome. We clustered the 1290 subsystems in the SEED database using four sequence

Table 1
Packages and software required to run SUPER-FOCUS

Software	Description	Version	Download	Required
Python	Programming language	>=2.6 and < 3	http://www.python.org/download	Yes
Jellyfish	*k*-mer counter	2.X	http://www.genome.umd.edu/jellyfish.html	No[a]
Numpy SciPy	Matrix manipulation	1.10.X 0.16.X	http://sourceforge.net/projects/numpy http://sourceforge.net/projects/scipy	Yes No[a]
RAPSearch2 DIAMOND BLAST	Homology search	2.X 0.7.X 2.X	http://rapsearch2.sourceforge.net https://github.com/bbuchfink/diamond ftp://ftp.ncbi.nlm.nih.gov/blast/executables/blast+/LATEST	Yes[b]

[a]Required only if the parameter "-focus 1" is used
[b]At least one of the aligners is required

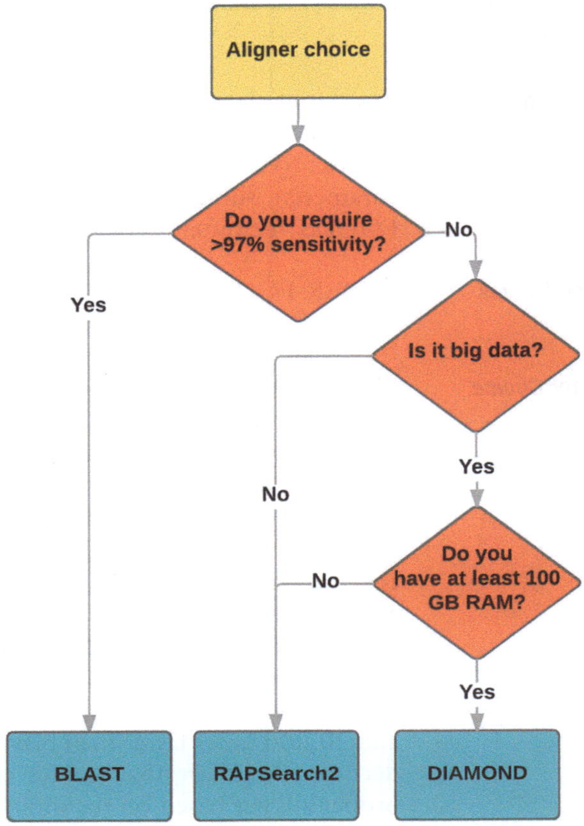

Fig. 1 Decision flowchart to help the user decide which aligner to use

identity cutoffs: 100, 98, 95, and 90%, creating four databases (DB_100, DB_98, DB_95, and DB_90). We showed [16] that DB_100 reduces ~6% of the total SEED database, DB_98 reduces ~30%, DB_95% reduces ~40%, and DB_90 reduces ~52% of the database. The more reduced the database means the less sensitive the result is; however, we showed in the SUPER-FOCUS manuscript that DB_90 has a loss of ~3% sensitivity. Therefore, the database should be chosen based on how sensitive the user wants the results.

2.5 Website

The latest version of SUPER-FOCUS can be found at http://edwards.sdsu.edu/superfocus. The website also includes a link to the program, a Q&A page, and contact information.

3 Methods

3.1 Downloading and Formatting the Database

After extracting files from the compressed file (.zip) downloaded from the latest SUPER-FOCUS version, it is necessary to format the database related to the aligner chosen. However, these two steps are done by one command that has three options (RAPSearch2, DIAMOND, and BLAST), each of which could be written together or separately. SUPER-FOCUS databases required ~85 GB of storage (~55 GB for RAPSearch2, ~11 GB for BLAST, ~10 GB for DIAMOND, and ~9 GB complete SEED). On prompt, inside the SUPER-FOCUS directory, the user will type the following command lines:

Format database for all aligners

```
$ python superfocus__downloadDB.py rapsearch blast diamond
```

Format database for a specific aligner

```
[1]$ python superfocus__downloadDB.py rapsearch
```

If the script ran perfectly, it will print the following message:

```
$ Done! Now you can run superfocus.py
```

3.2 Running SUPER-FOCUS

With the database downloaded and formatted, the user can run SUPER-FOCUS by following the basic command line:

```
$ python superfocus.py -q FASTA_or_FASTQ_File -dir myOutputdirectory -a alignerChoice[2]
```

Only the parameters **–q** and **–dir** are mandatory. The parameter **–q** is the pathway to query file (FASTA or FASTQ format). SUPER-FOCUS is also able to run multiples files in the same directory typing **–m 1** and in **–q** the pathway to the folder with multiple FASTA/FASTQ files. The parameter **–dir** is the pathway to an output directory.

All SUPER-FOCUS parameters are described in Table 2:

Examples

1. Performing the functional classification of one DNA fasta file, with DIAMOND database, and ten threads can be done using the following command:

```
$ python superfocus.py -q DNA_fasta_file -dir myOutput-directory -t 10 -a diamond
```

[1] This option can be one of rapsearch, blast, or diamond.

[2] Aligner choice [rapsearch (default), blast, or diamond]

Table 2
SUPER-FOCUS command line options

-h	———	Print help
-q	string	Query file (FASTA or FASTQ format) or folder with multiple FASTA/FASTQ files when -m 1
-m	int	Run the program for multiple files – 0 (False) / 1 (True) (default: 0)
-dir	string	Output directory
-o	string	Project name (default: 'my_project')
-mi	float	Minimum identity (%) (default: 60)
-ml	int	Minimum alignment (amino acids) (default: 15)
-t	int	Number of threads (default: 8)
-e	float	e-value (default: 0.00001)
-db	string	Database - DB_90 ; DB_95 ; DB_98; DB_100 (default: DB_98) . See "Database choice" for help on how to choose the database
-p	int	Amino acid input; 0 nucleotides; 1 amino acids (default: 0)
-a	string	aligner choice – rapsearch; blast; diamond (default: rapsearch)
-fast	int	runs RAPSearch2 or DIAMOND on fast mode - 0 (False) / 1 (True) (default: 1)
-n	int	Normalizes each query counts based on number of hits; 0 doesn't normalize; 1 normalizes (default: 1)
-focus	int	Predict taxa - 1 (True) / 0 (False) (default: 0)
-r	string	Use only the subsystems in the organisms predicted by "-focus"– ncbi/rast annotation (default: ncbi)

2. Performing the functional classification of 17 FASTQ files in the same directory, with BLAST database, three threads, and minimal identity of 80% can be done using the following command:

```
$ python superfocus.py -q DNA_fastq_files_directory -m
1 -dir myOutputdirectory  -t 3  -a blast -mi 80
```

3.3 Outputs

SUPER-FOCUS outputs four different files that contain the number of assignments and relative abundance of each label. The first file will contain only Subsystem Level 1 classification (more general)—e.g., Amino Acids and Derivatives—(Fig. 2). The second file will contain only Subsystem Level 2 classification—e.g., Aromatic amino acids and derivatives. The third file will contain only Subsystem Level 3 classification (more specialized) —e.g., Tryptophan synthesis. The last file will contain all information of the three last files (all levels) and SEED function information—e.g., Indole-3-glycerol phosphate synthase

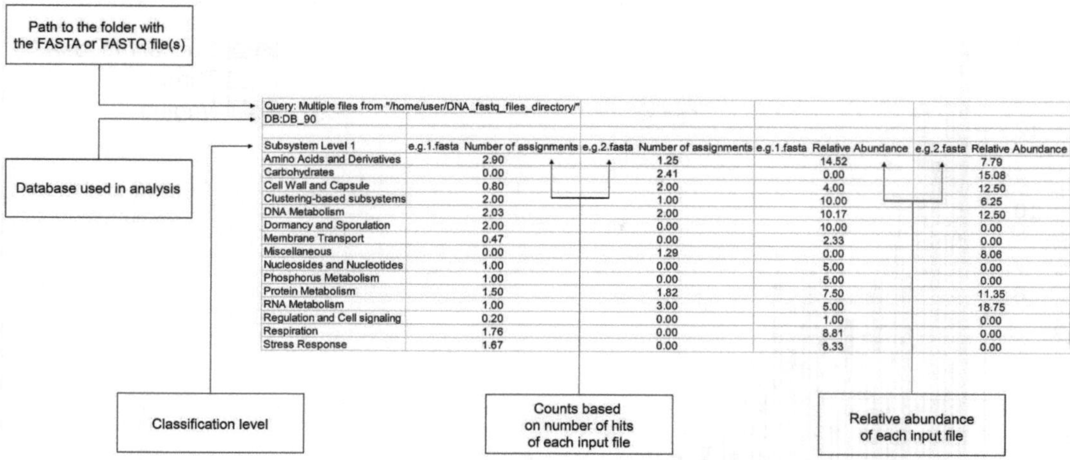

Fig. 2 SUPER-FOCUS output with subsystem Level 1 classification opened on spreadsheet software

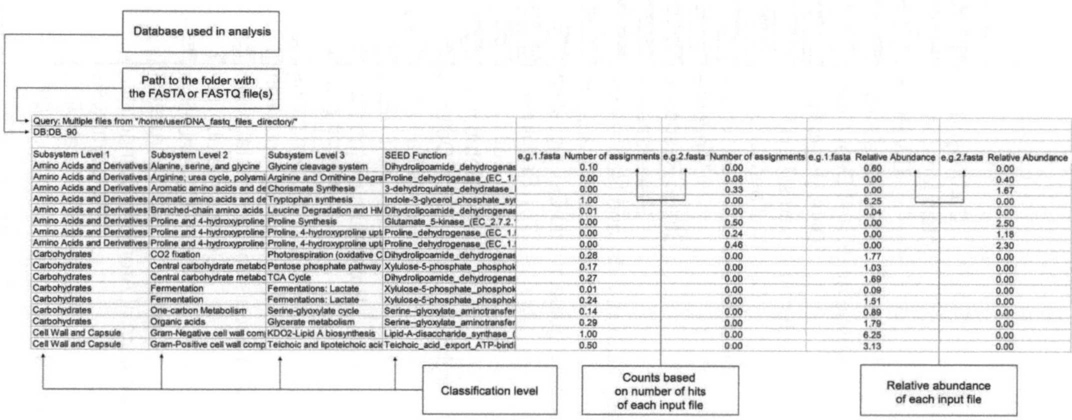

Fig. 3 SUPER-FOCUS output with all classification levels and SEED function opened on spreadsheet software. All levels and SEED function file

(EC 4.1.1.48)—(Fig. 3). All files created are .xls format and can be open easily in any spreadsheet software (e.g., LibreOffice Calc or Microsoft Excel).

3.4 Notes

1. DIAMOND and RAPSearch2 may not work properly using an already formatted database with different versions of the aligner. Therefore, the user should re-run "python superfocus__downloadDB.py AlignerChoice" whenever the DIAMOND/RAPSearch2 aligner was updated on the system.

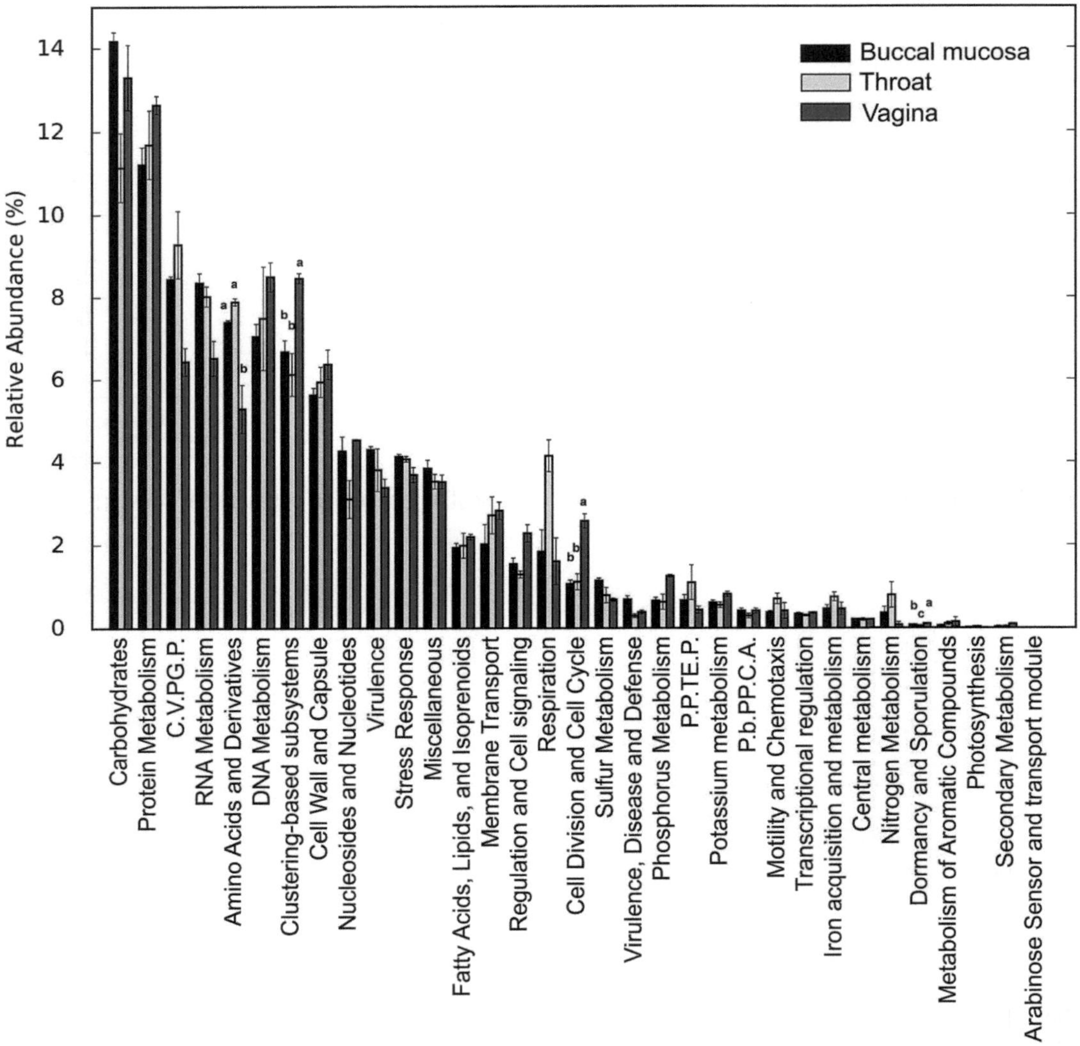

Fig. 4 Functional diversity of metagenome samples of Buccal mucosa (SRS023837, SRS015374, SRS016297), Throat (SRS065335, SRS014475, SRS015062), and Vaginal (SRS062752, SRS014465, SRS015071) from the HMP. Classification (relative abundance) was based on Level 1 in SUPER-FOCUS, and samples were compared using ANOVA ($p < 0.05$ Tukey-Kramer post-test ($\alpha = 0.05$), and Bonferroni correction using STAMP v.2.1.3. Different letters represent significant differences in relative abundance after post hoc tests. Abbreviations: C.V.PG.P. = Cofactors, Vitamins, Prosthetic Groups, Pigments; P.P.TE.P = Phages, Prophages, Transposable elements, Plasmids

2. It is not recommended to use SUPER-FOCUS on assembled data because long sequences (contigs) have many ORFs, and SUPER-FOCUS may classify only one of the ORFs. In this case, the user should first predict the ORFs, for example using FragGeneScan [19], and then run SUPER-FOCUS.

3. Only use DIAMOND for large datasets. It is slower than blastx for small datasets.
4. Joining paired-end reads guarantees a more accurate functional profiling and reduces the analysis run time. PEAR [20] and FLASH [21] can both join paired-end reads.
5. All the aligners should be installed in a directory listed in the Linux system path.

3.5 Case Studies

SUPER-FOCUS' output could be used for several applications such as visualizing differences among metagenomes. Here, we chose to work with metagenome samples from Buccal mucosa (SRS023837, SRS015374, SRS016297), Throat (SRS065335, SRS014475, SRS015062), and Vagina (SRS062752, SRS014465, SRS015071) from the human project (HMP) [3]. We used two output files as examples: Subsystem Level 1 and Level 3. In the first example, we demonstrate the differences between relative abundances (%) of each Level 1 classification (Fig. 4). The statistical test between metagenomes was performed using STAMP v.2.1.3 [22]. The metagenomes were compared by ANOVA ($p < 0.05$), Tukey-Kramer post-test ($\alpha = 0.05$), and Bonferroni correction. In the second example, we did a Principal Component Analysis (PCA) with relative abundances (%) from Level 3 file (Fig. 5).

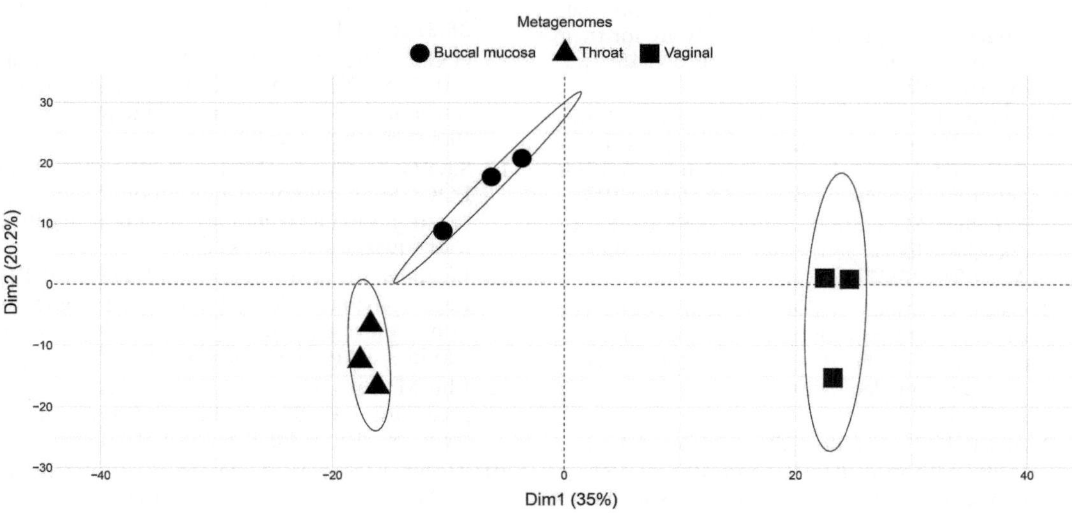

Fig. 5 Principal component analysis (PCA) of metagenome samples of Buccal mucosa (SRS023837, SRS015374, SRS016297), Throat (SRS065335, SRS014475, SRS015062), and Vaginal (SRS062752, SRS014465, SRS015071) from the HMP. Classification (relative abundance) was based on Level 3 in SUPER-FOCUS

Acknowledgments

We thank the SEED curators Dr Ross Overbeek, Dr Veronika Vonstein, and Dr Ramy Aziz for the amazing work on the annotation of subsystems since 2004. GGZS was supported by NSF Grants (CNS-1305112, MCB-1330800, and DUE-132809 to RAE and), and FACL was supported by Coordenação de Aperfeiçoamento de Pessoal de Nível Superior (CAPES/Brazil) fellowship.

References

1. Handelsman J (2004) Metagenomics: application of genomics to uncultured microorganisms. Microbiol Mol Biol Rev 68:669–685
2. Zhang J, Chiodini R, Badr A et al (2011) The impact of next-generation sequencing on genomics. J Genet Genomics 38:95–109
3. T.H.M.P. Consortium (2012) Structure, function and diversity of the healthy human microbiome. Nature 486:207–214
4. Sunagawa S, Coelho LP, Chaffron S et al (2015) Structure and function of the global ocean microbiome. Science 348:1261359
5. Mendoza MLZ, Sicheritz-Pontén T, Gilbert MTP (2015) Environmental genes and genomes: understanding the differences and challenges in the approaches and software for their analyses. Brief Bioinform 16(5):745–758. doi:10.1093/bib/bbv001
6. Overbeek R, Begley T, Butler RM et al (2005) The subsystems approach to genome annotation and its use in the project to annotate 1000 genomes. Nucleic Acids Res 33:5691–5702
7. Kanehisa M, Goto S (2000) KEGG: Kyoto encyclopedia of genes and genomes. Nucleic Acids Res 28:27–30
8. Caspi R, Altman T, Dale JM et al (2010) The MetaCyc database of metabolic pathways and enzymes and the BioCyc collection of pathway/genome databases. Nucleic Acids Res 38: D473–D479
9. Altschul SF, Madden TL, Schäffer AA et al (1997) Gapped BLAST and PSI-BLAST: a new generation of protein database search programs. Nucleic Acids Res 25:3389–3402
10. Kent WJ (2002) BLAT—The BLAST-like alignment tool. Genome Res 12:656–664
11. Zhao Y, Tang H, Ye Y (2012) RAPSearch2: a fast and memory-efficient protein similarity search tool for next-generation sequencing data. Bioinformatics 28:125–126
12. Huson DH, Beier S, Flade I, Górska A, El-Hadidi M, Mitra S, et al. MEGAN Community Edition—Interactive Exploration and Analysis of Large-Scale Microbiome Sequencing Data. PLOS Comput. Biol. 2016;12: e1004957
13. Mitra S, Rupek P, Richter DC et al (2011) Functional analysis of metagenomes and metatranscriptomes using SEED and KEGG. BMC Bioinformatics 12:S21
14. Meyer F, Paarmann D, D'Souza M et al (2008) The metagenomics RAST server—a public resource for the automatic phylogenetic and functional analysis of metagenomes. BMC Bioinformatics. 9:386
15. Edwards RA, Olson R, Disz T et al (2012) Real Time Metagenomics: Using k-mers to annotate metagenomes. Bioinformatics 28:3316–3317
16. G.G.Z. Silva, K.T. Green, B.E. Dutilh, et al. (2015) SUPER-FOCUS: a tool for agile functional analysis of shotgun metagenomic data, Bioinformatics. btv584
17. Silva GGZ, Cuevas DA, Dutilh BE et al (2014) FOCUS: an alignment-free model to identify organisms in metagenomes using non-negative least squares. PeerJ 2:e425
18. Berendzen J, Bruno WJ, Cohn JD et al (2012) Rapid phylogenetic and functional classification of short genomic fragments with signature peptides. BMC Res Notes 5:460
19. Rho M, Tang H, Ye Y (2010) FragGeneScan: predicting genes in short and error-prone reads. Nucleic Acids Res 38:e191–e191
20. Zhang J, Kobert K, Flouri T et al (2014) PEAR: a fast and accurate Illumina Paired-End reAd mergeR. Bioinformatics 30:614–620
21. Magoč T, Salzberg SL (2011) FLASH: fast length adjustment of short reads to improve genome assemblies. Bioinformatics 27:2957–2963
22. Parks DH, Tyson GW, Hugenholtz P et al (2014) STAMP: statistical analysis of taxonomic and functional profiles. Bioinformatics 30:3123–3124

Chapter 5

MPFit: Computational Tool for Predicting Moonlighting Proteins

Ishita Khan, Joshua McGraw, and Daisuke Kihara

Abstract

An increasing number of proteins have been found which are capable of performing two or more distinct functions. These proteins, known as moonlighting proteins, have drawn much attention recently as they may play critical roles in disease pathways and development. However, because moonlighting proteins are often found serendipitously, our understanding of moonlighting proteins is still quite limited. In order to lay the foundation for systematic moonlighting proteins studies, we developed MPFit, a software package for predicting moonlighting proteins from their omics features including protein–protein and gene interaction networks. Here, we describe and demonstrate the algorithm of MPFit, the idea behind it, and provide instruction for using the software.

Keywords Moonlighting proteins, Protein function prediction, Dual function, Function annotation, Protein association, Feature imputation, Omics-data, Genome

1 Introduction

While annotating gene function in a genome, the possibility that a gene has two or more distinct functions is usually not explicitly considered. However, an increasing number of proteins have been demonstrated to have more than one biological function, termed moonlighting proteins [1–3]. As long as the additional functions do not interfere with its primary function, moonlighting proteins (MPs) can benefit a cell in several ways. The existence of multifunctional proteins can aid in energy conservation during cell growth and reproduction as well as resulting in a more compact genome. Understanding the variety of MPs will also have clinical benefits as studies have identified a number of MPs that play important roles in cellular activities and biochemical pathways that are involved in cancer, metabolic disorders, and other diseases [4–7]. It has been suggested that for this reason the presence of MPs is under positive selection. This selective pressure and cell-level benefits of MPs

Daisuke Kihara (ed.), *Protein Function Prediction: Methods and Protocols*, Methods in Molecular Biology, vol. 1611,
DOI 10.1007/978-1-4939-7015-5_5,

suggest that moonlighting proteins in diverse genomes might be a common phenomenon.

Considering our current insufficient knowledge of moonlighting proteins, it is a significant challenge for computational protein function annotation to deal with moonlighting proteins. Conventional sequence-based functional annotations methods that are based on the concept of homology or conserved motifs/domains have difficulty identifying additional functions due to the existence of cases where homolog of a MP does not possess the secondary function [8] or has a different secondary function [9, 10]. A study by Gomez et al. compared 11 methods and reported that PSI-BLAST performed relatively well in identifying moonlighting functions [11]. Investigation on our function prediction tools, PFP [12–14] and ESG [15] in comparison with PSI-BLAST [16] has shown that PFP, which mines function information from weakly similar sequences, had the best performance in predicting two distinct functions of MPs [17]. These two studies suggest that secondary functions may be found within distantly related sequences if not among close homologs; however, because MPs are usually found unexpectedly by experiments, the datasets used in the two studies were limited and require further investigation. Due to the limited number of known MPs, systematic studies of MPs are still in its early stage for obtaining a comprehensive picture of proteins' moonlighting functions [18].

Until recently, there have been several bioinformatics approaches proposed for the detection of MPs, but they either relied heavily on the existence of functional annotation of a protein [19, 20] or addressed individual aspects of moonlighting proteins' functional diversity, i.e., sequence similarity [11, 17], motifs/domains [21], structural disorder [22], or protein-protein interaction (PPI) patterns combined with existing gene ontology annotations [19, 20, 23].

In contrast to these previous works, we have recently developed a computational framework for genome-scale characterization of MPs using comprehensive functional and context-based information of proteins [18]. Biological contexts examined in the study include networks of protein–protein interaction (PPI), similarity in the phylogenetic profile, gene expression profile correlation, and genetic interaction. We found that in general MPs tend to have more functionally diverse proteins in their networks, which would be reasonable considering the multi-functional nature of MPs.

Based on this study, we have constructed a prediction model named MPFit (Moonlighting Protein prediction with missing Feature imputation) for identifying moonlighting proteins [24]. To address the diverse nature of moonlighting proteins, MPFit uses various features of proteins ranging from gene ontology (GO) when available [25], several omics data, namely protein–protein interaction (PPI), gene expression, phylogenetic profiles, genetic interactions, and network-based graph properties (such as node between-ness, degree centrality, closeness-centrality), to protein structural properties such as the length of intrinsically disordered regions in the protein chain. For the omics features, interacting proteins to the target protein are clustered in terms of their functional similarity, and the numbers of clusters are used as features. In general, MPs have more clusters as they interact with proteins of diverse functions. For proteins that do not have certain features available in databases, we have additionally developed an imputation technique using random forest to predict missing features. These features are combined with machine learning classifiers to make moonlighting protein prediction.

MPFit was tested on a dataset of 268 known MPs from a manually curated database, MoonProt [26], which includes genomes such as human (45 proteins, 16.8%), *E. coli* (30 proteins, 11.19%), yeast (27 proteins, 10.1%), and mouse (11 proteins, 4.1%). The benchmark dataset also included 162 negative examples of MPs (termed as non-MP), which we computationally selected based on our previously established GO-based criteria [18]. The benchmark study on the dataset showed that MPFit predicted MPs with over 98% accuracy when proteins' GO terms were available. Using only non-GO-based features, MPFit maintained a high accuracy of over 75%. The latter result is important because it indicates that MPs can be identified by analyzing available omics data even without sufficient function annotations. Last, we have run MPFit with the best performing omics-based feature combinations on three genomes, *Saccharomyces cerevisiae* (yeast), *Caenorhabditis elegans*, and *Homo sapiens* (human) and found that about 2–10% of the proteomes are potential MPs.

2 The MPFit Algorithm

An overview of the MPFit algorithm is illustrated in Fig. 1. The top panel (Fig. 1a) shows the four phases that MPFit undergoes: feature data collection and construction, feature extraction, missing feature imputation (when needed), and classification for a query protein into moonlighting protein (MP) or non-moonlighting protein (non-MP). A broad range of features are used, i.e., GO annotations, PPI network, gene expression profiles (GE), phylogenetic profiles (Phylo), genetic interactions (GI), disordered protein regions (DOR), and the protein's graph properties in the PPI network (NET) (Fig. 1a).

Figure 1b shows the feature extraction phase for omics data-based features (i.e., PPI, GE, Phylo, and GI) for a protein P_i, we first build a network N_i for P_i. Each node in N_i is a protein; edges in N_i connect proteins that physically interact with each other (in the case of PPI) or that have significant correlation between

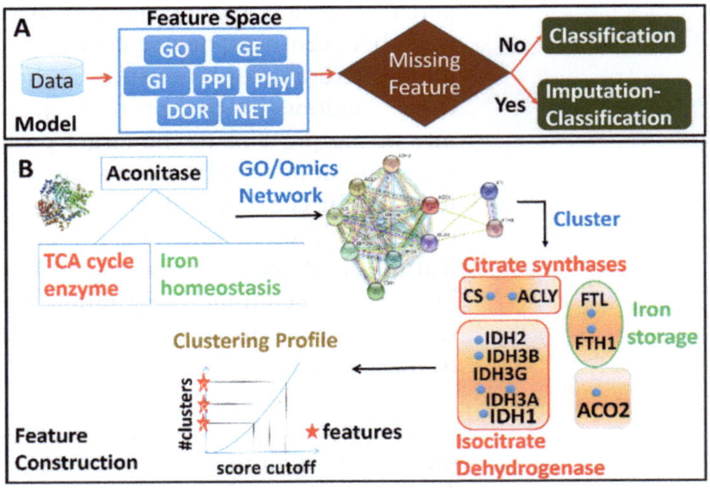

Fig. 1 Schematic diagram of MPFit. (**a**) Overall flowchart of the algorithm. (**b**) Feature construction. It shows the PPI feature extraction for human aconitase as an example

each other (for GE, Phylo, GI). Then proteins in the network are clustered with single linkage clustering in terms of their functional similarity based on their GO annotations. For clustering, several score thresholds are used and the number of clusters constructed at each threshold is recorded. Figure 1b illustrates the feature computation procedure for aconitase in human (*aco1*), an MP, for the PPI network. *Aco1* is an enzyme in the TCA cycle (the primary function) and is also involved in iron homeostasis (the second function). First, we extracted interacting partners for *aco1*, and then the PPI network was clustered based on the GO annotation similarity score of the interacting partners. The figure illustrates that four clusters of proteins were obtained for *aco1* using a certain GO similarity score threshold, where two of these clusters (circled in red) contain proteins related to the TCA cycle, while another cluster (green) was relevant to the second function. Such clustering was performed with five different similarity threshold scores (from 0.1 to 0.9 with an interval of 0.2), which resulted in a clustering profile shown in the bottom of Fig. 1b. Finally, we extracted the number of clusters at each score threshold as the PPI network features of *aco1*.

Proteins do not always have all of the feature data available that is used in MPFit. Thus MPFit uses a data imputation method based on the random-forest algorithm that fills in the missing features [27]. Figure 2 illustrates this procedure. Figure 2a shows the missing feature imputation process used when the MPFit algorithm was trained (i.e., parameters were optimized) using a training dataset and Fig. 2b shows how the imputation is applied in actual prediction.

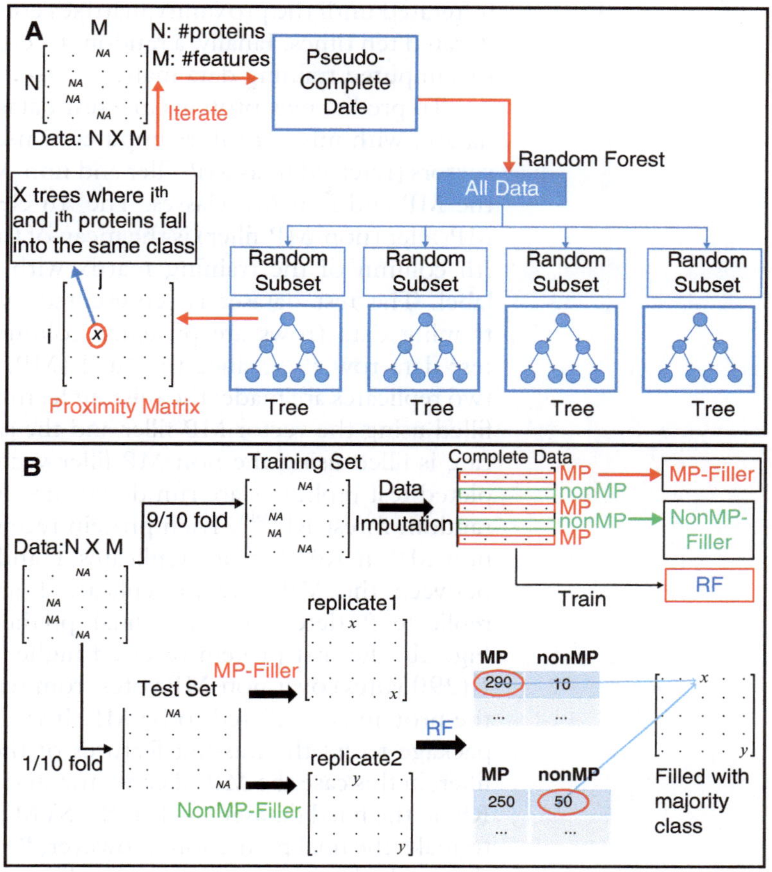

Fig. 2 Missing feature imputation in MPFit. (**a**) Feature imputation in the training stage. (**b**) Imputation in the testing stage

In Fig. 2a, the training dataset is represented top-left corner as a matrix where rows are proteins and columns are features. Missing features in the dataset are represented by NAs. The algorithm starts by replacing NAs with the column medians (i.e., median of the feature). Then a random forest was constructed using the feature set that are temporally filled by the previous step (pseudo-complete data in the figure). Random forest contains a number of decision trees, each of which is trained by a subset of the training set and feature combination, and thus in principle makes prediction in a different decision process. Using the random forest, each protein in the training set is predicted to be either a MP or non-MP, and the results are summarized in a so-called proximity matrix. The (i, j) element of the proximity matrix is the fraction of the trees in the random forest in which the proteins i and j fall in the same class. Now, the imputed value is updated to the weighted average of the non-missing features from other proteins, where weights are the proximities. When the missing features are determined, imputation is iterated until the proximity matrixes converge or the procedure is iterated ten times. Finally, a random forest RF^{train} is computed with this imputed training data matrix.

To predict new proteins in a test dataset (Fig. 2b), the training dataset with missing values imputed is used to compute two filler-vectors (referred to as MP-filler and non-MP-filler), one for each of the MP and non-MP classes. The *i*th element of the filler vector MP-filler (non-MP-filler) is the mean of the imputed features at the *i*th column of the training matrix with the MP (non-MP) class label. The test dataset is represented as a matrix similar to the training data (rows are proteins; columns are features). For the test data row r^{i}_{test}, since the label (MP/non-MP) is not known, two replicates are made; the missing features in the first replicate are filled using the vector MP-filler and the same for the second replicate is filled using the non-MP-filler vector. Now these two completed test replicates are run down through the previously trained random forest RF^{train}. Each protein receives tree votes of MP and non-MP in RF^{train} from replicates 1 and 2, and the higher vote between the MP vote in replicate 1 and the non-MP vote in replicate 2 determines the final prediction of the protein. In Fig. 2b, the first protein received higher MP votes from replicate 1 (290 votes) over non-MP votes from replicate 2 (50 votes); thus, the protein is predicted to be MP. It is also possible in the MPFit package to fill the missing features of the protein with the voted filler, in this case the MP-filler-vector, and run a classifier other than RF on the filled dataset (such as the SVM or naive-Bayes algorithm) to make the final prediction. However, RF is recommended as it has shown the best performance according to our original work.

3 Using MPFit

3.1 Installing MPFit

MPFit is made freely available for use and can be obtained from http://kiharalab.org/MPprediction. Under the Source Code section, select "MPFit Source Code" to download and extract the zipped archive. MPFit is written in Perl and requires three R package dependencies to run: randomForest, e1071, and stats. The package also includes example input files and output files.

3.2 MPFit Stages

Here, we provide a step-by-step description of the logical stages that the MPFit package undergoes for making a MP or non-MP prediction for query proteins.

Stage 1. Input file preparation. Input file to edit is

MPFit_generalized/Feature_Construction/Interacitons/input_uac.txt. Provide UniProt accession for each query protein, each on a new line. The UniProt ID is needed to retrieve feature data in the next stage.

Stage 2. Here the feature data of the queries is generated. MPFit uses following features: GO: Gene Ontology, PPI: Protein–Protein Interactions, Phylo: Phylogenetic profile, GE: Gene Expression, DOR: DisOrdered Regions, GI: Genetic Interactions, NET: 3 graph properties, i.e., between-ness, degree centrality, and closeness centrality of a node in PPI graph.

Data resources: The GO annotations for proteins were obtained from the UniProt database [28], and all GO annotations were extracted regardless of their GO evidence code (IEA or IDA). The STRING database [29] was used as the resource for the PPI network. Protein–protein physical association data were extracted from STRING for PPI. To construct the gene expression (GE) network, expression profiles were obtained from the COEXPRESdb database [30]. Gene pairs that had an absolute value of their Pearson correlation of expression levels within the top 2% among all the pairs in the database were connected as edges in the network. The Phylogenetic profile (Phylo) network was constructed using STRING. A protein pair was connected in the network if they had a sufficient score (>0.7 as recommended by STRING) at "neighborhood," "co-occurrence," or "gene-fusion" in the STRING database. For the genetic interaction (GI) network, we used the BIOGRID database [31]. Gene pairs were extracted that had the "experiment type" listed as "genetic" to be associated in the GI network. For the NET feature, three graph properties of proteins, namely, degree centrality, closeness centrality, and between-ness centrality, were computed from the PPI network. For the DOR feature, disordered region data are taken from the D2P2 database [32] and three properties were computed, namely, the number and the total length of disordered regions as well as the proportion of disordered regions in the entire protein sequence.

Run `$ get_interactions.pl` within each of the /GO (Gene Ontology), /PPI (Protein–protein interactions), /Phylo (Phylogenetic Profiles), and /GI (Genetic Interactions) directories of MPFit_generalized/Feature_Construction/Interactions. No actions are necessary for DOR (Disordered Protein Regions) and NET (Graph Properties in PPI) as the former is simply taken from the data resource and NET is computed from the PPI network later in Stage 3.

Stage 3. In this stage, features based on the data extracted in *Stage 2* are computed. The output of this stage is a feature file of the query proteins for each of the different feature spaces, with non-existing features represented as "NA," which will be imputed by the subsequent stage.

Run `$ compute_features <feature>` for each feature where <feature> is the type of feature to incorporate, i.e., GO, PPI, GI, Phylo, DOR, and NET. These are generated in MPFit_generalized/Feature_Construction/Construct/Features/<feature>.txt

Stage 4. Combine features and perform future imputation.

Change the directory to MPFit_generalized/MPFit_Model/ and run `$ combine_features.pl`

This will create feature files for the input dataset for all possible combinations of the seven omics-based features (PPI, GE, GI, Phylo, DOR, NET) in the MPFit_generalized/MPFit_Model/ Data/Features/directory. If the user wishes to omit usage of certain features, they can delete corresponding feature combination from the MPFit_generalized/MPFit_Model/Data/comb.txt flat file.

Stage 5. Run MPFit using a chosen classifier and obtain a prediction result for query proteins.

```
Run $ call_MP_impute_classify.pl  <classifier>
```

This command will perform imputation of missing features with all possible feature combinations mentioned above and run MPFit. Output of this stage is a MP/non-MP prediction of the query proteins, which is provided at MPFit_generalized/MPFit_Model/ Result/. Here, three possible classifiers can be given as input by the user: SVM (put 1 at <classifier>), naive Bayes (2), and random forest (3). Random forest is the recommended classifier to use as it showed improved performance over the others in our recent study [24]. Note that the MPFit prediction output is provided for all possible combinations of the seven omics-based features (PPI, GE, GI, Phylo, DOR, NET), and the GO feature. As we see in the next section, it is recommended to use a consensus of the predictions by two omics combinations to make a final MP/non-MP decision: Phylo+GE+GI+DOR+NET and PPI+Phylo+GE. According to the original paper [24], these two combinations performed well in the benchmark study in the case that the GO feature was not available for the query proteins. If the GO term feature is available, it is recommended to use a combination of GO and the two omics-based feature combination mentioned above: Phylo+GE+GI+DOR +NET or PPI+Phylo+GE. The resultant output file shows MP/ non-MP predictions for the input proteins, and the number of input proteins predicted as MP/non-MP (Fig. 3).

3.3 MPFit Prediction Accuracy

In the original work of MPFit [24], we reported prediction performance by using various different combinations of features, and concluded that two feature combinations, Phylo+GE+GI+DOR+NET and PPI+Phylo+GE, had high coverage (i.e., proteins in the benchmark dataset that had corresponding features and can be predicted) and high *F*-scores (0.796, 0.760 for coverage and 0.711, 0.754, for *F*-score, respectively). Thus, for the genome-scale prediction performed in the work, the consensus of the two was used. We have also tested three machine learning methods for the classifier in MPFit, random forest, naive Bayes, and SVM and reported that

Prediction Results using Phylo+GE+GI+DOR+NET

```
"-----------MPFit Output---------"
"P06745" "mp"
"P10809" "mp"
"P30041" "mp"
"Q43155" "mp"
"Q9CW03" "mp"
"Total Predicted MP"
5
"Total Predicted NONMP"
0
```

Prediction Results using PPI+Phylo+GE

```
"-----------MPFit Output---------"
"P06745" "mp"
"P10809" "mp"
"P30041" "mp"
"Q43155" "nonmp"
"Q9CW03" "mp"
"Total Predicted MP"
4
"Total Predicted NONMP"
1
```

Fig. 3 Example of the output file. Prediction of five query proteins, P06745, P10809, P30041, Q43155, and Q9CW03, is shown. These proteins are the same as those discussed in Tables 1 and 2. A prediction result using each feature combination is output in a separate file, but here we show results of two combinations, Phylo+GE+GI+DOR+NET and PPI+Phylo+GE, in one figure. "mp" or "nonmp" for a query protein indicates that it is predicted to be moonlighting or non-moonlighting proteins, respectively

random forest outperformed the other two. Hence, we recommend using the two best feature combinations (Phylo+GE+GI+DOR +NET or PPI+Phylo+GE) with random forest as the final classifier in the MPFit package.

3.4 Case Studies

Here, we show example predictions of five MPs and five non-MPs (Table 1). In Table 1, the first five proteins are MPs, while the latter five are non-MPs. Due to its dual-functional nature, GO term annotations of MPs are classified into a larger number of clusters than non-MPs' as shown in the rightmost column in Table 1. For the prediction, the two recommended feature combinations were used, Phylo+GE+GI+DOR+NET and PPI+Phylo+GE, with random forest as the final classifier. For the five MPs, as shown in Table 2, both feature combinations correctly predicted four proteins as MPs, P30041, P06745, Q9CW03, and P10809. For one

Table 1
Functions of moonlighting and non-moonlighting proteins used in the case study

Protein name	UniProt ID	Function 1 (related GO terms)[a]	Function 2 (related GO terms)	# of GO terms[b]	# of MF-GO Clusters at (0.1, 0.5) SS cutoff[c]
Structural maintenance of chromosomes protein 3	Q9CW03	Form a cohesion complex that maintains proper sister chromatid cohesion (GO:003603, GO:0030893)	Involved in the control of cell growth and transformation. (GO:0051301, GO:0019827)	30	(3, 7)
Ferredoxin-dependent glutamate synthase	Q43155	Glutamate synthase (GO:0006537, GO:0015930)	Subunit of UDP-sulfoquinovose synthase (GO:0051536, GO:0051538)	17	(3, 3)
Peroxiredoxin-6	P30041	Acidic calcium-independent Phospholipase (GO:0006629, GO:0009395)	Can reduce H_2O_2 and short chain organic, fatty acid, and phospholipid hydroperoxides (GO:0016491, GO:0051920)	22	(3, 4)
Glucose-6-phosphate isomerase	P06745	Catalyzes interconversion of glucose-6-phosphate and fructose-6-phosphate (GO:0004347, GO:0016853)	Binds to target cells and causes pre-B cells to mature into antibody secreting cells (GO:0008083, GO:0005125)	17	(2, 7)
60 kDa heat shock protein, mitochondrial	P10809	Protein chaperone, prevents proteins from misfolding, promotes correct refolding (GO:0006986, GO:0006457)	Receptor for HDL affinity for apolipoprotein apoA-II (GO:0002039, GO:0001530)	53	(2, 5)

Sec-independent protein translocase protein TatC	P69423	Transports large folded proteins (GO:0005887,GO:0005622)	-	8	(1, 1)
Probable aminoglycoside efflux pump	P24177	Participates in the efflux of aminoglycosides. (GO:0016021, GO:0016020)	-	8	(1, 1)
C4-dicarboxylic acid transporter DauA	P0AFR2	Responsible for the aerobic transport of succinate from the periplasm (GO:0005829,GO:0005887)	-	9	(1, 1)
PTS system glucose-specific EIICB component	P69786	catalyzes the phosphorylation of incoming sugar substrates (GO:0016021,GO:0005886)	-	10	(1, 1)
Low-affinity inorganic phosphate transporter 1	P0AFJ7	Low-affinity inorganic phosphate transport (GO:0005737, GO:0005886)	-	10	(1, 1)

The first five are moonlighting proteins and the latter five are non-moonlighting proteins

[a]In the columns for function 1 and function 2, representative GO terms related to the functions are shown. Non-moonlighting proteins do not have the secondary function, and thus the function 2 column is left as empty (-)

[b]# of GO terms, the number of GO terms in UniProt

[c]The numbers of clusters of the GO terms constructed by using Semantic functional Similarity (SS) score cutoff of 0.1 and 0.5 are shown. Moonlighting proteins tend to have more clusters, which indicate that they have more diverse GO term annotations reflecting their dual functions

Table 2
Prediction results for the five moonlighting proteins

UniProt ID	Q9CW03	Q43155	P30041	P06745	P10809
Phylo+GE+GI+DOR+NET/Random Forest	No	Yes	Yes	Yes	Yes
PPI+Phylo+GE/Random Forest	Yes	No	Yes	Yes	Yes

Table 3
Prediction results for the five non-moonlighting proteins

UniProt ID	P69423	P24177	P0AFR2	P69786	P0AFJ7
Phylo+GE+GI+DOR+NET/Random Forest	No	No	No	Yes	Yes
PPI+Phylo+GE/Random Forest	No	No	No	Yes	No

protein, Q43155, the five feature combination (Phylo+GE+GI +DOR+NET) predicted correctly as MPs, but the other did not. Table 3 shows the results for the five non-MPs. Three proteins were correctly predicted by the two feature sets as non-MPs; P0AFJ7 was predicted as non-MPs when PPI+Phylo+GE was used, but not when the other feature set was used. For P69786, both feature sets incorrectly predicted it as MPs.

Acknowledgments

This work was partly supported by the National Institute of General Medical Sciences of the National Institutes of Health (R01GM097528) and the National Science Foundation (IIS1319551, DBI1262189, IOS1127027, DMS1614777).

References

1. Campbell RM, Scanes CG (1995) Endocrine peptides 'moonlighting' as immune modulators: roles for somatostatin and GH-releasing factor. J Endocrinol 147(3):383–396
2. Jeffery CJ (1999) Moonlighting proteins. Trends Biochem Sci 24(1):8–11
3. Weaver DT (1998) Telomeres: moonlighting by DNA repair proteins. Curr Biol 8(14): R492–R494
4. Jeffery CJ (2011) Proteins with neomorphic moonlighting functions in disease. IUBMB Life 63(7):489–494
5. Jeffery CJ (2009) Moonlighting proteins—an update. Mol Biosyst 5(4):345–350
6. Jeffery CJ (2004) Molecular mechanisms for multitasking: recent crystal structures of moonlighting proteins. Curr Opin Struct Biol 14(6):663–668
7. Jeffery CJ (2003) Moonlighting proteins: old proteins learning new tricks. Trends Genet 19 (8):415–417
8. Ozimek P, Kotter P, Veenhuis M, van der Klei IJ (2006) Hansenula polymorpha and Saccharomyces cerevisiae Pex5p's recognize different, independent peroxisomal targeting signals in alcohol oxidase. FEBS Lett 580(1):46–50
9. Chen XJ, Wang X, Kaufman BA, Butow RA (2005) Aconitase couples metabolic regulation to mitochondrial DNA maintenance. Science 307(5710):714–717
10. Banerjee S, Nandyala AK, Raviprasad P, Ahmed N, Hasnain SE (2007) Iron-dependent RNA-

binding activity of Mycobacterium tuberculosis aconitase. J Bacteriol 189(11):4046–4052

11. Gomez A, Domedel N, Cedano J, Pinol J, Querol E (2003) Do current sequence analysis algorithms disclose multifunctional (moonlighting) proteins? Bioinformatics **19**(7):895–896
12. Hawkins T, Luban S, Kihara D (2006) Enhanced automated function prediction using distantly related sequences and contextual association by PFP. Protein Sci 15(6):1550–1556
13. Khan IK, Wei Q, Chitale M, Kihara D (2015) PFP/ESG: automated protein function prediction servers enhanced with Gene Ontology visualization tool. Bioinformatics 31(2):271–272
14. Hawkins T, Chitale M, Luban S, Kihara D (2009) PFP: Automated prediction of gene ontology functional annotations with confidence scores using protein sequence data. Proteins 74(3):566–582
15. Chitale M, Hawkins T, Park C, Kihara D (2009) ESG: extended similarity group method for automated protein function prediction. Bioinformatics 25(14):1739–1745
16. Altschul SF, Madden TL, Schaffer AA, Zhang J, Zhang Z, Miller W, Lipman DJ (1997) Gapped BLAST and PSI-BLAST: a new generation of protein database search programs. Nucleic Acids Res 25(17):3389–3402
17. Khan I, Chitale M, Rayon C, Kihara D (2012) Evaluation of function predictions by PFP, ESG,and PSI-BLAST for moonlighting proteins. BMC Proc 6(Suppl 7):S5
18. Khan I, Chen Y, Dong T, Hong X, Takeuchi R, Mori H, Kihara D (2014) Genome-scale identification and characterization of moonlighting proteins. Biol Direct 9:30
19. Chapple CE, Robisson B, Spinelli L, Guien C, Becker E, Brun C (2015) Extreme multifunctional proteins identified from a human protein interaction network. Nat Commun 6:7412
20. Pritykin Y, Ghersi D, Singh M (2015) Genome-wide detection and analysis of multifunctional genes. PLoS Comput Biol 11(10): e1004467
21. Hernandez S, Franco L, Calvo A, Ferragut G, Hermoso A, Amela I, Gomez A, Querol E, Cedano J (2015) Bioinformatics and moonlighting proteins. Front Bioeng Biotechnol 3:90
22. Hernandez S, Amela I, Cedano J, Pinol J, Perez-Pons J, Mozo-Villarias A, Querol E (2012) Do moonlighting proteins belong to the intrinsically disordered protein class? J Proteomics Bioinform 5:262–264
23. Gomez A, Hernandez S, Amela I, Pinol J, Cedano J, Querol E (2011) Do protein–protein interaction databases identify moonlighting proteins? Mol Biosyst 7(8):2379–2382
24. Khan IK, Kihara D (2016) Genome-scale prediction of moonlighting proteins using diverse protein association information. Bioinformatics 32(15):2281–2288
25. Consortium GO (2015) Gene Ontology Consortium: going forward. Nucleic Acids Res 43 (Database issue):D1049–D1056
26. Mani M, Chen C, Amblee V, Liu H, Mathur T, Zwicke G, Zabad S, Patel B, Thakkar J, Jeffery CJ (2015) MoonProt: a database for proteins that are known to moonlight. Nucleic Acids Res 43(Database issue):D277–D282
27. Breiman L (2001) Random forests. Mach Learn 45(1):5–32
28. UniProt Consortium (2015) UniProt: a hub for protein information. Nucleic Acids Res 43 (Database issue):D204–D212
29. Szklarczyk D, Franceschini A, Wyder S, Forslund K, Heller D, Huerta-Cepas J, Simonovic M, Roth A, Santos A, Tsafou KP, Kuhn M, Bork P, Jensen LJ, von Mering C (2015) STRING v10: protein-protein interaction networks, integrated over the tree of life. Nucleic Acids Res 43(Database issue):D447–D452
30. Okamura Y, Aoki Y, Obayashi T, Tadaka S, Ito S, Narise T, Kinoshita K (2015) COXPRESdb in 2015: coexpression database for animal species by DNA-microarray and RNAseq-based expression data with multiple quality assessment systems. Nucleic Acids Res 43(Database issue):D82–D86
31. Oughtred R, Chatr-aryamontri A, Breitkreutz BJ, Chang CS, Rust JM, Theesfeld CL, Heinicke S, Breitkreutz A, Chen D, Hirschman J, Kolas N, Livstone MS, Nixon J, O'Donnell L, Ramage L, Winter A, Reguly T, Sellam A, Stark C, Boucher L, Dolinski K, Tyers M (2016) BioGRID: a resource for studying biological interactions in yeast. Cold Spring Harb Protoc (1):pdb top080754
32. Oates ME, Romero P, Ishida T, Ghalwash M, Mizianty MJ, Xue B, Dosztanyi Z, Uversky VN, Obradovic Z, Kurgan L, Dunker AK, Gough J (2013) D(2)P(2): database of disordered protein predictions. Nucleic Acids Res 41(Database issue):D508–D516

Chapter 6

Predicting Secretory Proteins with SignalP

Henrik Nielsen

Abstract

SignalP is the currently most widely used program for prediction of signal peptides from amino acid sequences. Proteins with signal peptides are targeted to the secretory pathway, but are not necessarily secreted. After a brief introduction to the biology of signal peptides and the history of signal peptide prediction, this chapter will describe all the options of the current version of SignalP and the details of the output from the program. The chapter includes a case study where the scores of SignalP were used in a novel way to predict the functional effects of amino acid substitutions in signal peptides.

Keywords Signal peptides, Prediction, Secretion, Protein sorting, Protein subcellular location

1 Introduction

A signal peptide (SP) is the N-terminal part of a protein that is targeted to the secretory pathway in both pro- and eukaryotes [1] (*see*, however, **Note 1**). In eukaryotes, a protein with an SP will be targeted to the endoplasmic reticulum (ER) membrane and be co-translationally translocated across the membrane. In prokaryotes, translocation takes place across the cytoplasmic membrane (inner membrane in Gram-negative bacteria), and the process can happen during or after translation. The SP-carrying protein is threaded through a protein complex known as the translocon, comprising the subunits SecY, E, and G in bacteria and Sec61 α, β, and γ in eukaryotes [2]. During translocation, the SP is cleaved off by an enzyme known as signal peptidase I or leader peptidase (*Lep*) in bacteria or signal peptidase complex in eukaryotes [3]. *See* **Notes 2–4** for exceptions to this general picture.

It is important to stress that the presence of an SP does *not* necessarily mean that the protein is secreted to the extracellular environment—it only means that it enters the secretory pathway. In all kinds of organisms, the protein could have one or more trans-membrane helices downstream of the SP and therefore be retained in the membrane [4]. In eukaryotes, the protein could also be

Daisuke Kihara (ed.), *Protein Function Prediction: Methods and Protocols*, Methods in Molecular Biology, vol. 1611,
DOI 10.1007/978-1-4939-7015-5_6, © Springer Science+Business Media LLC 2017

retained in one of the compartments that belong to the secretory pathway: the ER, the Golgi apparatus, or the lysosome/vacuole [5]; or it could be anchored to the outer face of the cytoplasmic membrane by a glycophosphatidylinositol (GPI) group [6]. In Gram-negative bacteria, the protein could be retained in the periplasm, or be inserted into the outer membrane as a β-barrel transmembrane protein [7]. In Gram-positive bacteria, the protein could be attached to the cell wall [8].

SPs are generally described as having three regions: an N-terminal n-region of variable length characterized by positive charge, a central h-region of at least 7 hydrophobic residues, and a C-terminal c-region of typically 3–7 polar residues. Positions −1 and −3 relative to the cleavage site are occupied by small uncharged residues; in bacteria predominantly Alanine. SPs of Gram-positive bacteria tend to be longer than those of Gram-negative bacteria, which in turn tend to be longer than eukaryotic SPs [1].

The SP is among the earliest prediction targets for bioinformatic algorithms, with the first simple prediction methods being published already in the 1980s [9–11]. In the early 1990s, a few machine learning methods were published [12, 13], but SignalP version 1.0 [14, 15] was in 1996 the first machine learning method for SP prediction to be made into a publically available web server. SignalP 1.0 and 1.1 were based on artificial neural networks (ANNs), while SignalP 2.0 from 1998 [16] added a hidden Markov model (HMM) prediction to better distinguish between SPs and signal anchors (transmembrane helices close to the N-terminus). SignalP 3.0 from 2004 [17] introduced the *D*-score for better discrimination between SPs and other sequences and retained the HMM option, while SignalP 4.0 from 2011 [18] is again purely ANN-based. While constructing SignalP 4.0, we did retrain the HMM part, but we found that it did not perform better than the ANNs in any of the performance parameters we tested. The most important new feature of SignalP 4.0 is the improved discrimination between signal peptides and transmembrane regions.

SignalP was updated to version 4.1 in 2012 with an option to set the *D*-score cutoff values so that the sensitivity is the same as that of SignalP 3.0, and an option to set the minimum cleavage site position in the sequence (the minimum SP length). More details about these options are given in Subheading 3.1. In addition, the documentation on the website was completely rewritten, and a FAQ was added.

Earlier versions of SignalP have repeatedly been reported as the best performing method in independent benchmarks [19–22]. SignalP 4 has not yet been independently evaluated, but in the SignalP 4.0 paper [18] we compared the performance to ten other methods and found that it was superior. The best competing methods were the combined SP and transmembrane helix predictors Phobius [23], Philius [24], and SPOCTOPUS [25].

Interestingly, the advantage of SignalP 4.0 over these three programs was larger for bacteria than for eukaryotes. This may be due to the fact that these three methods did not divide their training data into different organism groups but pooled them all together, resulting in methods that are optimized for the most abundant organism group in the data, the eukaryotes.

The performance values for SignalP 3.0 and 4.0 and the ten competing methods can be found in Table E of the supplementary materials of the SignalP 4.0 paper, which is available on the SignalP web site (click on "Article abstracts" and then "Update to SignalP v. 4.0"). It should be noted that those values are calculated by cross-validation on a homology-reduced data set, i.e., they are the performances you should expect when submitting proteins that are unrelated to anything in the SignalP 4.0 data set. When submitting close homologs to proteins in the SignalP 4.0 data set, a higher performance should be expected (compare the aforementioned Table E with the table on the "performance" page of the website documentation).

2 Materials

1. *Input data:* Amino acid sequences in FASTA format. Note that any letters not corresponding to the 20 standard amino acids, e.g., "U," "B," or "Z," will be converted to "X" and treated as unknown amino acids. *See* also **Notes 5** and **6**.
2. *Website:* SignalP 4.1 is available at http://www.cbs.dtu.dk/services/SignalP/, *see* Fig. 1. The previous versions are also kept online; just click "version history" near the top of the page.
3. *Downloadable package:* For those who prefer running SignalP on their own computers, there is an option to download a software package for command line use. The package is free for academic institutions, while there is a license fee for commercial users. Academic users can go to the page http://www.cbs.dtu.dk/cgi-bin/nph-sw_request?signalp to fill out the details and accept the license, while commercial users are asked to contact software@cbs.dtu.dk. The package is available for Linux, IRIX, Darwin (Mac OS X), and from March 2016 also for Windows computers via the free Unix-like environments provided by Cygwin [26] or MobaXterm [27].

3 Methods

Running SignalP with the default options is straightforward: On the website, you paste or upload the sequences and click "submit"; on the command line you write "signalp *input.fasta*." The output

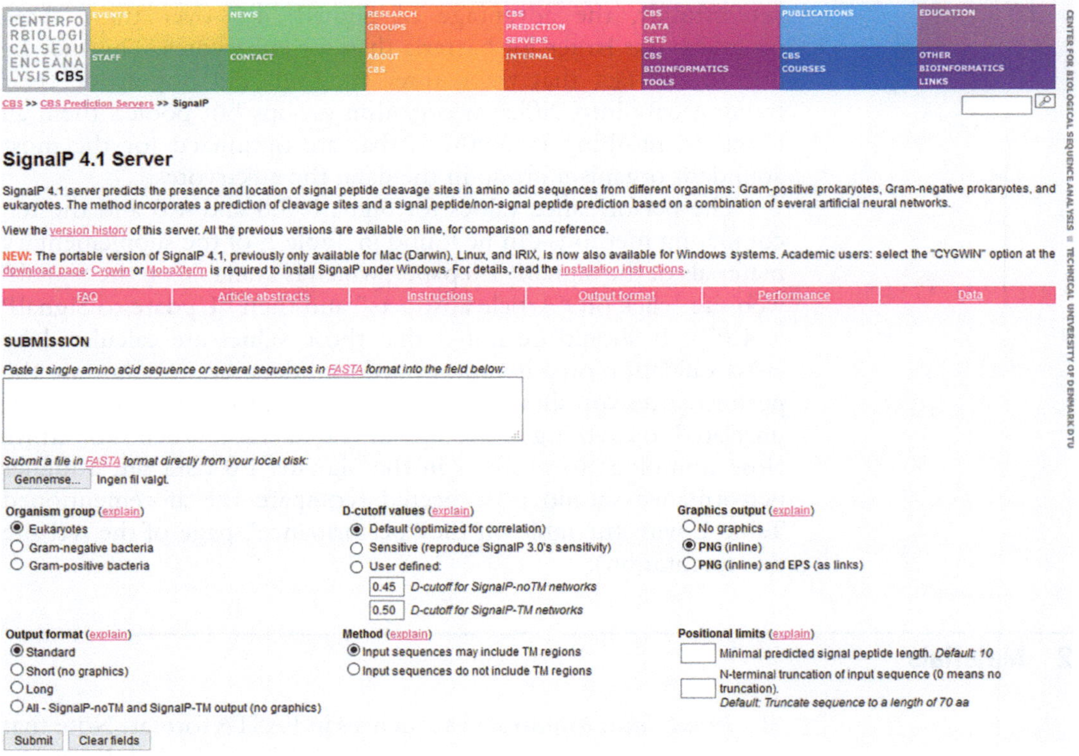

Fig. 1 The SignalP 4.1 web site, showing the input field and the available options

will tell you, for each sequence, whether there is an SP predicted, and if yes, where the cleavage site is predicted to be. However, as seen in Fig. 1, there are a number of options, of which especially "*Organism group*" and "*Cutoff*" are important to know about, and there are details of the output format that will help interpret the predictions.

3.1 Options

3.1.1 Organism Group

It is important to choose the correct organism group—Eukaryotes, Gram-negative bacteria, or Gram-positive bacteria—otherwise, the predictive performance will suffer. In this context, Gram-positive bacteria are defined as the phyla Actinobacteria (high G + C Gram-positive bacteria) and Firmicutes. Gram-negative bacteria are defined as all bacteria having both a plasma membrane and an outer membrane—basically all other bacteria except for the phylum Tenericutes (*Mycoplasma* and related genera). SignalP probably should not be used for Tenericutes at all, since they seem to lack a type I signal peptidase completely [28]. On the command line, organism group is chosen with one of the options "-t euk" (the default), "-t gram−", or "-t gram+." Concerning organism groups, *see* also **Notes 7–11**.

3.1.2 Output Format

There are four levels of detail possible: "short," "standard," "long," and "all." The two first formats report scores and conclusion at the sequence level; "short" in a one-line format and "standard" in a more human-readable format. "Standard" is the default on the web server, and "short" on the command line. The "long" and "all" formats additionally report scores for each position in each sequence (for an explanation of the scores, *see* Subheading 3.2). The difference between "long" and "all" is that "long" reports scores for the chosen ANN method only, while "all" reports scores for both ANN methods (SignalP-noTM and SignalP-TM, *see* Subheading 3.1.4 below for an explanation). On the command line, output format is chosen with the "-f" option; note that "standard" is chosen with "-f summary."

3.1.3 Graphics Output

SignalP can make a plot of the scores for each position in each sequence in portable network graphics (PNG) format and optionally also in encapsulated postscript (EPS) format. The default on the web is to make PNG graphics, while the default on the command line is no graphics. If you want graphics from the command line, use the options "-g png" or "-g png+eps."

3.1.4 Method

SignalP 4 has two sets of ANNs: SignalP-noTM is trained with only cytosolic and nuclear proteins in the negative set, while SignalP-TM is trained with a negative set that also included transmembrane proteins. During training, we found that the two methods SignalP-TM and SignalP-noTM were to some extent complementary, i.e., SignalP-TM did not yield as good results as SignalP-noTM when there were no transmembrane sequences involved. As a compromise, SignalP 4 per default uses a heuristic to decide which of the two sets of networks is used for the prediction of each sequence. If the user is positive that all proteins in the input are soluble, it is possible to override this heuristic and get a slightly better performance by using only the SignalP-noTM networks. This is done in the web interface by selecting "Input sequences do not include TM regions" and on the command line by including the option "-s notm."

3.1.5 Cutoff

The *D*-score (*see* Subheading 3.2) is used for determining whether each input sequence contains an SP or not. The user can set cutoff values (for SignalP-TM and SignalP-noTM separately) if a different balance between sensitivity and specificity is desired. The web interface offers two sets of predefined cutoff values, "Default" and "Sensitive." The "Default" cutoffs, corresponding to SignalP 4.0, are optimized to give the best Matthews correlation coefficient (*see* the "Performance" page on the website for definition), but they result in a quite conservative prediction with a sensitivity that is actually lower than that of SignalP 3.0. The "Sensitive" cutoffs, introduced in SignalP 4.1, are set to reproduce the sensitivity of

SignalP 3.0. This of course results in a slightly higher false-positive rate, but still significantly better than that of SignalP 3.0 when measured on the whole data set (with transmembrane proteins included in the negative set). Our recommendation is to use the "Sensitive" setting if it is important to avoid false negatives, but use the "Default" setting for estimating the proportion of SPs in an organism. The estimation by the "Default" cutoff was found to be in accordance with an estimate of the number of SPs in *Escherichia coli* by a recent proteogenomics study [29]. At the website, you can see the preset cutoff values change when you select "Default" or "Sensitive" or change the organism group. On the command line, the "Sensitive" cutoffs are selected by including the options "-U 0.34 -u 0.34" for organism group "euk" or "-U 0.42 -u 0.42" for organism groups "gram+" and "gram−."

3.1.6 Truncation of Input Sequence

By default, SignalP truncates every sequence to 70 amino acids before prediction. This gives enough included sequence after the cleavage site to give the optimal prediction for the vast majority of SPs. If you want to predict extremely long signal peptides, you can try a higher value, or disable truncation completely by entering 0 (zero). Note that the neural networks are trained with sequences with a maximal length of 70, and they include the relative position in the sequence in their input. Therefore, general performance may deteriorate if you change this setting. On the command line, truncation is changed with the "-c" option.

3.1.7 Minimal Predicted Signal Peptide Length

SignalP 4.0 could, in rare cases, erroneously predict extremely short signal peptides. These errors have, in SignalP 4.1, been eliminated by imposing a lower limit on the cleavage site position (SP length). The minimum length is by default ten, but you can adjust it. Signal peptides shorter than 15 residues are very rare, at the time of writing there are 17 experimentally confirmed cases in UniProt that are not fragments. If you want to disable this length restriction completely, enter 0 (zero). On the command line, minimal SP length is changed with the "-M" option.

3.2 Output

The neural networks in SignalP produce three output scores for each position in the input sequence:

1. *C-score* (raw cleavage site score): The output from the cleavage site networks, which are trained to distinguish SP cleavage sites from everything else. Note the position numbering of the cleavage site: The *C*-score is trained to be high at the position immediately *after* the cleavage site (the first residue in the mature protein).
2. *S-score* (signal peptide score): The output from the signal peptide networks, which are trained to distinguish positions within

SPs from positions in the mature part of the proteins and from proteins without SPs.

3. *Y-score* (combined cleavage site score): A combination (geometric average) of the *C*-score and the slope of the *S*-score, resulting in a better cleavage site prediction than the raw *C*-score alone. This is due to the fact that multiple high-peaking *C*-scores can be found in one sequence, where only one is the true cleavage site. The *Y*-score distinguishes between *C*-score peaks by choosing the one where the slope of the *S*-score is steep.

The graphical output from SignalP (Figs. 2 and 3) shows the three different scores, *C*, *S*, and *Y*, for each position in the sequence.

In the summary below the plot, the maximal values of the three scores are reported. In addition, the following two scores are shown:

1. *mean S:* The average *S*-score of the possible SP (from position 1 to the position immediately before the maximal *Y*-score).

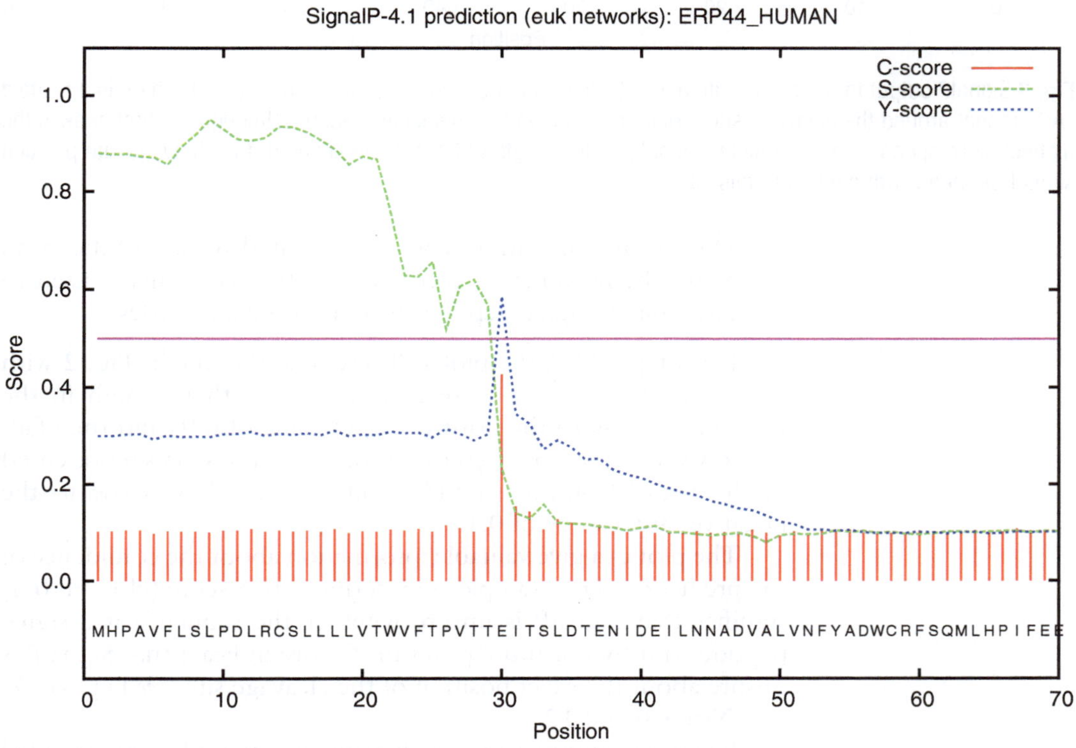

Fig. 2 SignalP output for a protein with a typical signal peptide, Human endoplasmic reticulum resident protein 44. Note that there is one conspicuous peak in *Y*-score at position 30, meaning that the signal peptide is predicted to be cleaved between amino acids 29 and 30. Please note that this protein is not secreted

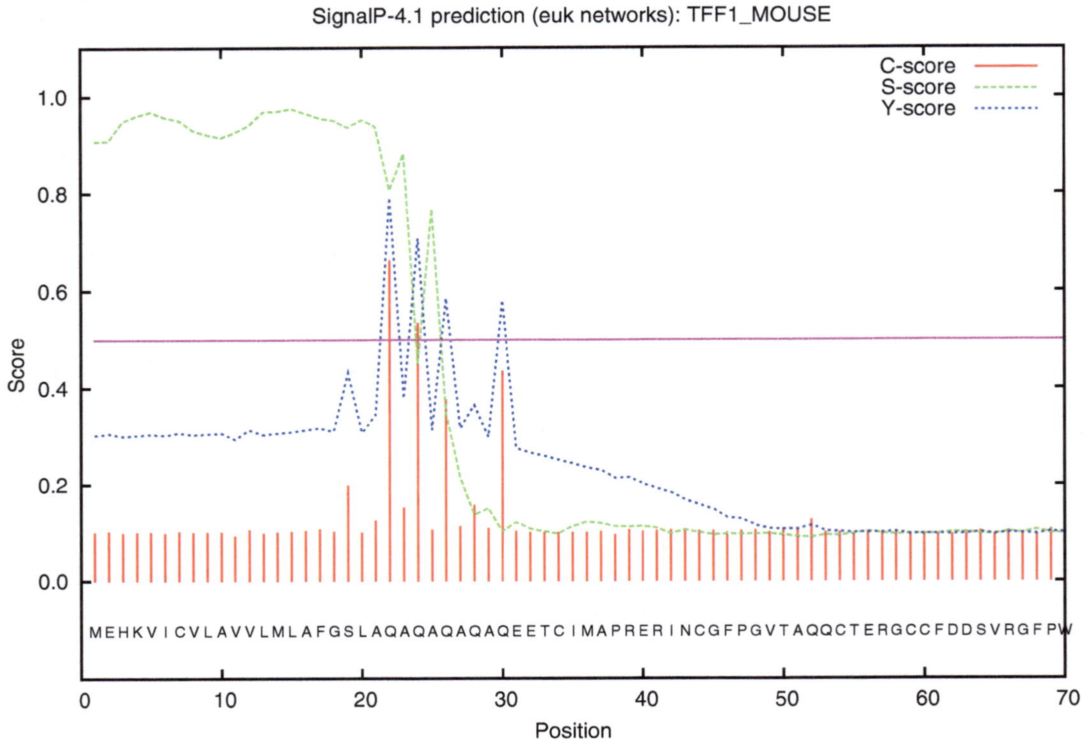

Fig. 3 SignalP output for a protein with a less typical signal peptide, Mouse Trefoil factor 1. There is a fivefold "AQ" repeat around the cleavage site region, resulting in four peaks in *Y*-score. Although the first peak is the highest, corresponding to a predicted signal peptide length of 21, the prediction of the cleavage site position should be taken with caution in this case

2. *D-score* (discrimination score): A weighted average of the mean *S* and the maximal *Y* scores. This is the score that is used to discriminate signal peptides from nonsignal peptides.

For a typical SP, the plot will resemble the one in Fig. 2 with one peak in *C*- and *Y*-score and an *S*-score that is high in the beginning (close to the positive target value of 0.9) and then falls to a low value. For non-secretory proteins all the scores represented in the SignalP output should ideally be very low (close to the negative target value of 0.1).

The plot can give valuable information about the confidence of the prediction. For example, an intermediate *S*-score (close to 0.5) signifies that SignalP is unsure whether the sequence is a signal peptide, and two or more peaks in *Y*-score indicate that SignalP is unsure about the exact position of the cleavage site (*see* Fig. 3). *See* also **Notes 6** and **12**.

Below the summary for each sequence, two files are provided via links: "data" and "gnuplot script." If you have the free graphics program gnuplot [30] on your computer, you can use these two files to customize your plot. If you want to keep these files when using the command line interface, you need to include the option "-k."

Below the output for all the sequences, two other files are provided via links, if at least one SP has been predicted. These are "processed fasta entries," a FASTA sequence file containing the sequences of those proteins that had predicted SPs, with the SP removed; and "gff file of processed entries," a file showing the signal peptides feature of those proteins that had predicted SPs in the format GFF (gene-finding format). Note that these two files are not produced by default in the command line interface; if you want them, include the options "-m *filename*" for processed FASTA entries or "-n *filename*" for GFF entries. *See*, however, **Note 13**.

The file with processed FASTA entries can be very useful for downstream analysis of those proteins that were predicted to have SPs. For example, if the focus is on predicting secreted proteins, it should always be checked whether there are predicted transmembrane helices downstream of the SP. This can, e.g., be done by submitting the processed FASTA entries to TMHMM [31, 32]. The advantage of using the processed FASTA entries instead of the entire sequences is that you get rid of the false-positive transmembrane helix predictions that TMHMM often makes for SPs.

3.3 Case Study

Since 1996, SignalP has been used to predict countless SPs. The three most influential papers about SignalP [14, 17, 18] have been cited more than 11,000 times in total according to Web of Science [33]. It is difficult to single out one particular SP prediction study as more interesting than the others.

However, one study from a group at the company Genentech [34] deserves special mention, since they used the output scores of SignalP in a creative way we had not anticipated. The paper is from 2009 and is therefore based on SignalP 3.0, but the same approach should be applicable to version 4.1. The focus of the study is the prediction of the functional effects of amino acid substitutions in SPs. It is important to stress that SignalP was not designed for this purpose—SignalP has been trained on wild-type sequences only, and mutated SPs that have lost their function partially or completely are more similar to wild-type SPs than to wild-type non-SPs. Mutated SPs can therefore be said to occupy a different part of sequence space than the wild-type SPs and non-SPs comprising the SignalP training set, and the task of predicting consequences of amino acid substitutions in SPs is probably a harder problem than the one for which SignalP was designed.

Predicting whether amino acid substitutions (non-synonymous single nucleotide polymorphisms, nsSNPs) have functional consequences for proteins is an intensely studied problem in bioinformatics. In general, the problem is defined as the discrimination between a positive data set of known disease-causing mutations and a negative data set of presumed neutral nsSNPs based on the amino acid sequence and the pattern of conservation around each substituted amino acid. One of the first and best known predictors for this problem is SIFT (sorting intolerant from tolerant) [35–37].

The Genentech authors gathered data sets of disease-causing mutations and neutral nsSNPs which occurred within the signal peptide region of human secretory proteins, and they hypothesized that the disease-causing mutations interfered with signal peptide function and therefore should be predictable by SignalP. They then defined a novel score based on the "long" format output of SignalP. This was termed the "*R*-score" and defined thus:

$$R = \max(\Delta S_{1,\ldots,n}) - \min(\Delta S_{1,\ldots,n}) + \max(\Delta C_{1,\ldots,n}) - \min(\Delta C_{1,\ldots,n})$$

where ΔS_i is the difference in *S*-score between the mutant and the wild type at position *i*, and $\max(\Delta S_{1,\ldots,n})$ is the maximal value of that difference within the entire predicted SP (*n* is the cleavage site position predicted for the wild type). The same definitions apply to the terms with *C*-score.

The authors were able to show that the *R*-score was significantly better than the simple difference in *D*-score (ΔD) for discriminating between disease-causing mutations and neutral nsSNPs. The performance of the *R*-score was similar to that of the score from SIFT which does no SP prediction. Furthermore, they showed that *R*-score and SIFT-score were not correlated, suggesting that they contributed independent information about the discrimination. Accordingly, a simple combination of *R*- and SIFT-score gave a better discrimination than either score alone.

It could be interesting to see whether these results are still valid when using SignalP 4.1 and some of the newer alternatives to SIFT, such as PolyPhen-2 [38], PON-P2 [39], SNAP2 [40], etc. In addition, more advanced ways of combining SignalP output with the outputs of other programs may prove to perform even better. A Chinese group in 2012 used a Random Forest classifier coupled with a feature selection scheme to integrate SignalP 3.0 output with sequence profiles and physicochemical parameters [41]. Their approach is interesting, and they report a large increase in discrimination performance relative to the *R*-score; but their result may be marked by overfitting, since no effort has been done to avoid homology between training and test sets. Homology reduction has been a crucial element in the construction of the SignalP datasets ever since version 1.0 [42] and homology partitioning is used in recent nsSNP effect predictors [39, 40].

A closely related question is whether SignalP is able to predict functional consequences of artificially induced mutations in SPs. We have tried using both *D*-score and *R*-score on a set of mutated

SPs from bacteria, but found the effects surprisingly difficult to predict (results not published). One possible future direction for SignalP could be to use such data in the training phase to improve the prediction of effects of mutations, whether naturally occurring or artificially induced. A method trained in this way might also yield improved predictions of secretion efficiency, cf. **Note 12**.

4 Notes

1. *SP definition:* Sometimes, especially in introductory textbooks, the term "signal peptide" has been used in a broader sense, meaning any (or any cleavable) sorting signal embedded in the amino acid sequence of a protein. However, the definition of SP given in Subheading 1 corresponds to the usage in most of the scientific literature, as well as in UniProt [43], Wikipedia [44], and the Sequence feature ontology [45]. Signals for import into mitochondria and chloroplasts are properly termed *transit peptides* and can be predicted, e.g., with the program TargetP [46, 47].
2. *Uncleaved SPs:* There are rare examples of SPs that are not cleaved (at the time of writing, there are 79 such cases annotated in UniProt). These should not be confused with *signal anchors*, which are transmembrane helices close to the N-terminus. Uncleaved SPs are very differently predicted by SignalP; some look like typical SPs, some look like typical non-SPs, and others have high *S*-scores but low *C*- and *Y*-scores.
3. *Bacterial lipoproteins:* Bacterial lipoproteins have special signal peptides that are cleaved by signal peptidase II, also known as Lipoprotein signal peptidase (*Lsp*). A diacylglyceryl group is attached to a conserved Cys residue in position +1 relative to the cleavage site, which bears no resemblance to the signal peptidase I cleavage site [48]. SignalP often predicts such sequences as SPs, but with a wrong cleavage site. For the prediction of prokaryotic lipoproteins, we recommend using the LipoP server [49, 50].
4. *Tat signal peptides:* The special SPs that direct some bacterial proteins through the Tat (Twin arginine translocation) pathway instead of the Sec pathway are not very well predicted by SignalP. These SPs have a special motif containing two Arginines in the N-terminal part, and they are in general longer and less hydrophobic than normal SPs [51, 52]. For the prediction of Tat signal peptides, we recommend using the TatP server [53, 54].

5. *Nucleotide sequences:* Note that SignalP will *not* produce sensible output from DNA sequences; it will treat such sequences as proteins exclusively consisting of Ala, Cys, Gly, and Thr.
6. *Start codon prediction:* Since SignalP predicts an N-terminal signal, it is dependent on a correct start codon assignment. A start codon assigned too far downstream will cut part of the true amino acid sequence, while a start codon assigned too far upstream will add arbitrary sequence to the N-terminus, both making it difficult to recognize a possible SP. If you get a prediction of an unusually long SP where the *S*-score is low in the beginning but then rises to a higher value, you should look for possible alternative start codons downstream of the annotated one. For eukaryotic sequences, you might want to check start codon predictions with the program NetStart [55, 56].
7. *Archaea:* There is no "Archaea" option among the organism groups. This is certainly not because Archaea do not have SPs, but because there are too few experimentally confirmed SPs from Archaea (at the time of writing, the number is 11 in UniProt).
8. *Viruses:* There is no "Viruses" option among the organism groups. Virus or phage SPs should be predicted according to their host organism group.
9. *Atypical Gram-positives:* Certain bacteria, notably *Deinococcus* spp., have a thick cell wall and react positively to the Gram-staining procedure, even though they also have an outer membrane [57]. Their SPs should probably be predicted with "Gram-negative bacteria" as organism group, although too few SPs from such organisms are experimentally known to answer this question confidently.
10. *SP diversity within eukaryotes:* It has long been known that some yeast signal peptides are not recognized by mammalian cells [58]. Therefore, it would be natural to assume that separate SignalP versions for yeast and Mammalia would provide better predictions than a common eukaryotic version. While developing SignalP 4.0 we tried dividing the eukaryotic data into animals, fungi, and plants and training separate methods for these three groups. However, this did not give any improvement, and performance for all three groups was better when using the method trained on all eukaryotic sequences together. This should be tested again for the next version of SignalP.
11. *SP diversity within bacteria:* The Gram-negative version of SignalP is probably biased toward *E. coli* and other γ-proteobacteria, since these constitute the bulk of the experimentally annotated Gram-negative SPs in UniProt. Newer results suggest that some bacteria have rather divergent cleavage site

motifs [59]. Future versions of SignalP might therefore benefit from dividing the Gram-negative bacteria into several groups, if enough data are available.

12. *Secretion efficiency:* A frequently asked question is whether SignalP can predict secretion efficiency when attaching an SP to a heterologous protein—in other words, whether efficient SPs score higher than slowly secreting SPs. The answer is unfortunately not known. Intuitively, one would expect efficient secretors to have higher *Y*- and *D*-scores. However, SignalP is trained to recognize SPs against a background of non-SPs, regardless of secretion efficiency. This means that inefficient secretors are trained with the same target value as efficient secretors, as long as they are naturally occurring SPs. Therefore, the scores will not necessarily correlate with secretion efficiency.
13. *GFF version:* The gene-finding format you get with the "-n" option is GFF version 2, which is actually deprecated [60]. This should be updated to GFF version 3 in the next SignalP version.

Acknowledgments

Heartfelt thanks go to all coauthors on the SignalP papers though the years: Søren Brunak, Jacob Engelbrecht, Gunnar von Heijne, Anders Krogh, Jannick Dyrløv Bendtsen, and Thomas Nordahl Petersen. In addition, I wish to thank the people who helped in implementing the website and still work on keeping it up and running: Kristoffer Rapacki, Hans Henrik Stærfeldt, and Peter Wad Sackett.

References

1. von Heijne G (1990) The signal peptide. J Membr Biol 115:195–201. doi:10.1007/BF01868635
2. Pohlschröder M, Prinz WA, Hartmann E, Beckwith J (1997) Protein translocation in the three domains of life: variations on a theme. Cell 91:563–566. doi:10.1016/S0092-8674(00)80443-2
3. Dalbey RE, Lively MO, Bron S, Dijl JMV (1997) The chemistry and enzymology of the type I signal peptidases. Protein Sci 6:1129–1138. doi:10.1002/pro.5560060601
4. von Heijne G (1988) Transcending the impenetrable: how proteins come to terms with membranes. Biochim Biophys Acta Rev Biomembr 947:307–333. doi:10.1016/0304-4157(88)90013-5
5. Harter C, Wieland F (1996) The secretory pathway: mechanisms of protein sorting and transport. Biochim Biophys Acta Rev Biomembr 1286:75–93. doi:10.1016/0304-4157(96)00003-2
6. Ferguson MAJ, Williams AF (1988) Cell-surface anchoring of proteins via Glycosyl-phosphatidylinositol structures. Annu Rev Biochem 57:285–320. doi:10.1146/annurev.bi.57.070188.001441
7. Duong F, Eichler J, Price A et al (1997) Biogenesis of the gram-negative bacterial envelope. Cell 91:567–573. doi:10.1016/S0092-8674(00)80444-4
8. Mazmanian SK, Liu G, Ton-That H, Schneewind O (1999) *Staphylococcus aureus* Sortase, an enzyme that anchors surface proteins to the

Cell Wall. Science 285:760–763. doi:10.1126/science.285.5428.760

9. von Heijne G (1983) Patterns of amino acids near signal-sequence cleavage sites. Eur J Biochem 133:17–21. doi:10.1111/j.1432-1033.1983.tb07424.x
10. McGeoch DJ (1985) On the predictive recognition of signal peptide sequences. Virus Res 3:271–286. doi:10.1016/0168-1702(85)90051-6
11. von Heijne G (1986) A new method for predicting signal sequence cleavage sites. Nucleic Acids Res 14:4683–4690. doi:10.1093/nar/14.11.4683
12. Ladunga I, Czakó F, Csabai I, Geszti T (1991) Improving signal peptide prediction accuracy by simulated neural network. Comput Appl Biosci 7:485–487. doi:10.1093/bioinformatics/7.4.485
13. Schneider G, Wrede P (1993) Development of artificial neural filters for pattern recognition in protein sequences. J Mol Evol 36:586–595. doi:10.1007/BF00556363
14. Nielsen H, Brunak S, Engelbrecht J, von Heijne G (1997) Identification of prokaryotic and eukaryotic signal peptides and prediction of their cleavage sites. Protein Eng 10:1–6. doi:10.1093/protein/10.1.1
15. Nielsen H, Engelbrecht J, Brunak S, Heijne GV (1997) A neural network method for identification of prokaryotic and eukaryotic signal peptides and prediction of their cleavage sites. Int J Neural Syst 8:581–599. doi:10.1142/S0129065797000537
16. Nielsen H, Krogh A (1998) Prediction of signal peptides and signal anchors by a hidden Markov model. Proc Int Conf Intell Syst Mol Biol 6:122–130
17. Bendtsen JD, Nielsen H, von Heijne G, Brunak S (2004) Improved prediction of signal peptides: SignalP 3.0. J Mol Biol 340:783–795. doi:10.1016/j.jmb.2004.05.028
18. Petersen TN, Brunak S, von Heijne G, Nielsen H (2011) SignalP 4.0: discriminating signal peptides from transmembrane regions. Nat Methods 8:785–786. doi:10.1038/nmeth.1701
19. Menne KML, Hermjakob H, Apweiler R (2000) A comparison of signal sequence prediction methods using a test set of signal peptides. Bioinformatics 16:741–742. doi:10.1093/bioinformatics/16.8.741
20. Klee E, Ellis L (2005) Evaluating eukaryotic secreted protein prediction. BMC Bioinformatics 6:1–7. doi:10.1186/1471-2105-6-256
21. Choo K, Tan T, Ranganathan S (2009) A comprehensive assessment of N-terminal signal peptides prediction methods. BMC Bioinformatics 10:S2. doi:10.1186/1471-2105-10-S15-S2
22. Zhang X, Li Y, Li Y (2009) Evaluating signal peptide prediction methods for gram-positive bacteria. Biologia (Bratisl) 64:655–659. doi:10.2478/s11756-009-0118-3
23. Käll L, Krogh A, Sonnhammer EL (2004) A combined transmembrane topology and signal peptide prediction method. J Mol Biol 338:1027–1036. doi:10.1016/j.jmb.2004.03.016
24. Reynolds SM, Käll L, Riffle ME et al (2008) Transmembrane topology and signal peptide prediction using dynamic Bayesian networks. PLoS Comput Biol 4:e1000213. doi:10.1371/journal.pcbi.1000213
25. Viklund H, Bernsel A, Skwark M, Elofsson A (2008) SPOCTOPUS: a combined predictor of signal peptides and membrane protein topology. Bioinformatics 24:2928–2929. doi:10.1093/bioinformatics/btn550
26. Cygwin. https://cygwin.com/. Accessed 30 May 2016
27. MobaXterm free Xserver and tabbed SSH client for Windows. http://mobaxterm.mobatek.net/. Accessed 30 May 2016
28. Fraser CM, Gocayne JD, White O et al (1995) The minimal gene complement of *Mycoplasma genitalium*. Science 270:397–404. doi:10.1126/science.270.5235.397
29. Ivankov DN, Payne SH, Galperin MY et al (2013) How many signal peptides are there in bacteria? Environ Microbiol 15:983–990. doi:10.1111/1462-2920.12105
30. gnuplot homepage. http://www.gnuplot.info/. Accessed 30 May 2016
31. Krogh A, Larsson B, von Heijne G, Sonnhammer EL (2001) Predicting transmembrane protein topology with a hidden Markov model: application to complete genomes. J Mol Biol 305:567–580. doi:10.1006/jmbi.2000.4315
32. TMHMM Server, v. 2.0. http://www.cbs.dtu.dk/services/TMHMM/. Accessed 30 May 2016
33. Henrik Nielsen D-4128-2011—ResearcherID.com. http://www.researcherid.com/rid/D-4128-2011. Accessed 30 May 2016
34. Hon LS, Zhang Y, Kaminker JS, Zhang Z (2009) Computational prediction of the functional effects of amino acid substitutions in signal peptides using a model-based approach. Hum Mutat 30:99–106. doi:10.1002/humu.20798

35. Ng PC, Henikoff S (2003) SIFT: predicting amino acid changes that affect protein function. Nucleic Acids Res 31:3812–3814. doi:10.1093/nar/gkg509
36. Vaser R, Adusumalli S, Leng SN et al (2016) SIFT missense predictions for genomes. Nat Protoc 11:1–9. doi:10.1038/nprot.2015.123
37. SIFT—Predict effects of nonsynonmous/missense variants. http://sift.bii.a-star.edu.sg/. Accessed 30 May 2016
38. Adzhubei IA, Schmidt S, Peshkin L et al (2010) A method and server for predicting damaging missense mutations. Nat Methods 7:248–249. doi:10.1038/nmeth0410-248
39. Niroula A, Urolagin S, Vihinen M (2015) PON-P2: prediction method for fast and reliable identification of harmful variants. PLoS One 10:e0117380. doi:10.1371/journal.pone.0117380
40. Hecht M, Bromberg Y, Rost B (2015) Better prediction of functional effects for sequence variants. BMC Genomics 16:S1. doi:10.1186/1471-2164-16-S8-S1
41. Qin W, Li Y, Li J et al (2012) Predicting deleterious non-synonymous single nucleotide polymorphisms in signal peptides based on hybrid sequence attributes. Comput Biol Chem 36:31–35. doi:10.1016/j.compbiolchem.2011.12.001
42. Nielsen H, Engelbrecht J, von Heijne G, Brunak S (1996) Defining a similarity threshold for a functional protein sequence pattern: the signal peptide cleavage site. Proteins Struct Funct Bioinforma 24:165–177. doi:10.1002/(SICI)1097-0134(199602)24:2<165::AID-PROT4>3.0.CO;2-I
43. UniProt help: Signal peptide. http://www.uniprot.org/help/signal. Accessed 30 May 2016
44. Signal peptide—Wikipedia, the free encyclopedia. https://en.wikipedia.org/wiki/Signal_peptide. Accessed 30 May 2016
45. SO_0000418 < Ontology Lookup Service < EMBL-EBI. http://www.ebi.ac.uk/ols/ontologies/so/terms?iri=http%3A%2F%2Fpurl.obolibrary.org%2Fobo%2FSO_0000418. Accessed 30 May 2016
46. Emanuelsson O, Nielsen H, Brunak S, von Heijne G (2000) Predicting subcellular localization of proteins based on their N-terminal amino acid sequence. J Mol Biol 300:1005–1016. doi:10.1006/jmbi.2000.3903
47. TargetP 1.1 Server. http://www.cbs.dtu.dk/services/TargetP/. Accessed 30 May 2016
48. von Heijne G (1989) The structure of signal peptides from bacterial lipoproteins. Protein Eng 2:531–534. doi:10.1093/protein/2.7.531
49. Juncker AS, Willenbrock H, von Heijne G et al (2003) Prediction of lipoprotein signal peptides in gram-negative bacteria. Protein Sci 12:1652–1662. doi:10.1110/ps.0303703
50. LipoP 1.0 Server. http://www.cbs.dtu.dk/services/LipoP/. Accessed 30 May 2016
51. Cristóbal S, de Gier J-W, Nielsen H, von Heijne G (1999) Competition between Sec- and TAT-dependent protein translocation in *Escherichia coli*. EMBO J 18:2982–2990. doi:10.1093/emboj/18.11.2982
52. Berks BC (2015) The twin-arginine protein translocation pathway. Annu Rev Biochem 84:843–864. doi:10.1146/annurev-biochem-060614-034251
53. Bendtsen JD, Nielsen H, Widdick D et al (2005) Prediction of twin-arginine signal peptides. BMC Bioinformatics 6:167. doi:10.1186/1471-2105-6-167
54. TatP 1.0 Server. http://www.cbs.dtu.dk/services/TatP/. Accessed 30 May 2016
55. Pedersen AG, Nielsen H (1997) Neural network prediction of translation initiation sites in eukaryotes: perspectives for EST and genome analysis. Proc Int Conf Intell Syst Mol Biol 5:226–233
56. NetStart 1.0 Prediction Server. http://www.cbs.dtu.dk/services/NetStart/. Accessed 30 May 2016
57. Thompson BG, Murray RGE (1981) Isolation and characterization of the plasma membrane and the outer membrane of *Deinococcus radiodurans* strain Sark. Can J Microbiol 27:729–734. doi:10.1139/m81-111
58. Bird P, Gething MJ, Sambrook J (1987) Translocation in yeast and mammalian cells: not all signal sequences are functionally equivalent. J Cell Biol 105:2905–2914. doi:10.1083/jcb.105.6.2905
59. Payne SH, Bonissone S, Wu S et al (2012) Unexpected diversity of signal peptides in prokaryotes. mBio 3:e00339–e00312. doi:10.1128/mBio.00339-12
60. GFF2—GMOD. http://gmod.org/wiki/GFF2. Accessed 30 May 2016

Chapter 7

The ProFunc Function Prediction Server

Roman A. Laskowski

Abstract

The ProFunc web server is a tool for helping identify the function of a given protein whose 3D coordinates have been experimentally determined or homology modeled. It uses a cocktail of both sequence- and structure-based methods to identify matches to other proteins that may, in turn, suggest the query protein's most likely function. The server was originally developed to aid the worldwide structural genomics effort at the start of the millennium. It accepts a file containing the protein's 3D coordinates in PDB format, and, when processing is complete, sends an email containing a link to the password-protected result pages. The results include an at-a-glance summary, as well as separate pages containing more detailed analyses. The server can be found at: http://www.ebi.ac.uk/thornton-srv/databases/profunc.

Keywords ProFunc, Function prediction, Protein 3D structure, Structural templates, Motifs, Fold matching

1 Introduction

The years 2000–2015 saw a rapid development and refinement of high-throughput methods for protein structure determination. The aim of these structure determination pipelines was to solve as many protein 3D structures as possible, specifically those from families where no structural information was available, or which had a specific biological interest. Over 13,500 structures were determined during this period [1] and made a significant contribution to structural biology and the biomedical sciences.

One of the problems with this flood of new structures, however, was that the function of about a third of the proteins was unknown, and for about half it was designated as "putative" [2]. This threatened to seriously undermine the value of these structural models, so the structural genomics centers undertook to develop novel tools for helping identify the likely function of a protein from its 3D coordinates. These included tools that used sequence-comparison methods only, such as the Fold and Function Assignment Server (FFAS) [3], or structure-based methods, such as

Daisuke Kihara (ed.), *Protein Function Prediction: Methods and Protocols*, Methods in Molecular Biology, vol. 1611, DOI 10.1007/978-1-4939-7015-5_7, © Springer Science+Business Media LLC 2017

the pocket-comparison method pvSOAR [4]. Two tools made use of both sequence- and structure-based prediction methods: ProKnow [5] and ProFunc [6]. In this chapter we describe the use of the latter.

ProFunc is a web server that uses existing and novel methods to analyze the submitted protein's sequence and 3D structure to identify functional motifs or close relationships to functionally characterized proteins. The benefit of using multiple methods is that, if one method fails to find anything useful, another might provide a meaningful match. Furthermore, if several complementary methods indicate the same or similar function, the confidence in that prediction is enhanced. The system is easy to use, requiring only the submission of a PDB-format file and an email address where the link to the results should be sent.

In addition to matches to other proteins, the system provides more local matches, which can be particularly useful when the protein's active site is of interest. The most informative matches tend to be those returned by the "residue template searches," described later. These identify local structural matches (encompassing a sphere of radius 10 Å) and, as we shall see later, can provide a very informative functional match. They can pinpoint similarities as well as differences in the protein's binding site which might account for, say, functional differences across different members of the same protein family.

In addition to the ProFunc results, the system also generates a full set of PDBsum analyses for the submitted structure. PDBsum is a pictorial atlas of PDB structures, with a page for each entry in the PDB [7, 8]. It provides a number of schematic diagrams including secondary structure assignments, topology diagrams, protein–ligand, protein–DNA and protein–protein interactions, as well as visualizing clefts in the protein's structure.

2 Materials

2.1 Input Data

As mentioned above, the primary input to ProFunc is just a PDB-format file of the protein's 3D coordinates. Also required is an email address for the return of a link to the results pages. A large number of processes are run on the submitted protein and can take several hours if there are many submissions in the queue. Usually, the results are returned within 1–2 h.

The processes that are run on the submitted structure fall into three categories, as shown in Table 1: sequence searches, structure-based analyses, and 3D template methods.

Table 1
The sequence- and structure-based methods used in ProFunc

Sequence searches	Structure-based analyses	3D template methods
FASTA sequence search vs. PDB	Fold search using **PDBeFold**	Enzyme active sites
BLAST sequence search vs. UniProt	Surface clefts computed by **SURFNET**	Ligand binding sites
InterProScan seq motifs (PROSITE, BLOCKS, SMART, Pfam, etc.)	Residue conservation computed by **ScoreCons**	DNA-binding sites
	Nest analysis	Reverse templates

2.2 Programs Run by ProFunc

2.2.1 Sequence Searches

The sequence searches use standard methods. A FASTA search [9] scans the protein's sequence against those in the Protein Data Bank (PDB) [10] to find any matches to known 3D structures. A BLAST [11] search is then made against the protein sequences in the UniProt Knowledgebase [12] and the hits aligned using a simple pile-up procedure to give a multiple sequence alignment. This is passed to the ScoreCons program [13] to calculate conservation scores for each residue position.

Finally, InterProScan [14] identifies any sequence motifs, patterns, and fingerprints present in the query sequence. This uses a number of motif databases including PROSITE [15], Pfam [16], SMART [17], PRINTS [18], SUPERFAMILY [19], and ProDom [20].

2.2.2 Structure-Based Analyses

The structure-based analyses, which use the query protein's 3D coordinates, are listed in the middle column of Table 1. The first is a fold search against the structures in the PDB using PDBeFold (formerly called SSM for Secondary Structure Matching [21]) which uses a fast graph-matching algorithm to compare the secondary structure elements (SSEs) of the target structure against those of the structures in its database. Any strong matches are superposed and an r.m.s.d. for equivalent C^{α} atoms calculated. Proteins with similar folds often (but not always) perform similar functions.

Clefts in the query protein's structure are computed using SURFNET [22] as these often have functional significance—the largest is often the protein's binding site [23]. When allied with residue conservation they provide a particularly powerful means of identifying the protein's key residues.

The last of the structure-based analyses is an identification of any structural motifs called "nests" in the structure. A "nest" is an anion or cation binding site formed by three or more consecutive amino acids whose main-chain φ-ψ dihedral angles alternate between the right- and left-handed α and γ regions of the Ramachandran plot. Such motifs are found to be frequently associated with protein functional sites [24, 25].

2.2.3 3D Template Methods

For the final batch of searches, four different types of 3D templates are used: enzyme active sites, ligand binding sites, DNA-binding sites, and so-called reverse templates [26]. All consist of specific 3D conformations of between three and six amino acid residues. The first three types are compiled from structures in the PDB—the enzymes templates manually, and the ligand- and DNA-binding templates automatically—and are scanned against the query protein structure. The "reverse" templates are generated from the query structure itself before being scanned against a representative set of structures in the PDB.

All template searches are performed by a program called Jess [27] which can rapidly find residue conformations matching a given template. To significantly reduce the numbers of false positives returned, a second process compares the similarity of the local environments in the matched proteins by first pairing up residues in equivalent 3D positions within 10 Å of the template's geometrical center. A similarity score, and associated expectation value (E-value), are computed, as described later.

The 3D template searches, which focus on local similarity, can sometimes pick up even quite distant homologues whose sequences have markedly diverged over evolutionary time but whose functional sites have had to be conserved to retain their biological roles.

2.3 Website URLs

The ProFunc URL is: http://www.ebi.ac.uk/thornton-srv/databases/profunc

PDBsum is at: http://www.ebi.ac.uk/pdbsum

The results can be accessed via either site using the supplied four-character code and password. Both sites have documentation describing the outputs, while the ProFunc site also has a detailed tutorial (on which parts of this chapter have been based).

3 Methods

3.1 ProFunc Home Page

The ProFunc home page is where you can upload your PDB file, request an analysis of an existing entry in the PDB, or locate the results of a previous run. Figure 1a shows the initial entry form.

1. The first option, **A**, is for uploading your own protein structure coordinates—which must be in valid PDB format (*see* **Note 1**). Just click the **Browse** button, locate the file on your computer and click **Upload**. You will then be asked for your name, institute, and email address (Fig. 1b). If the PDB file contains

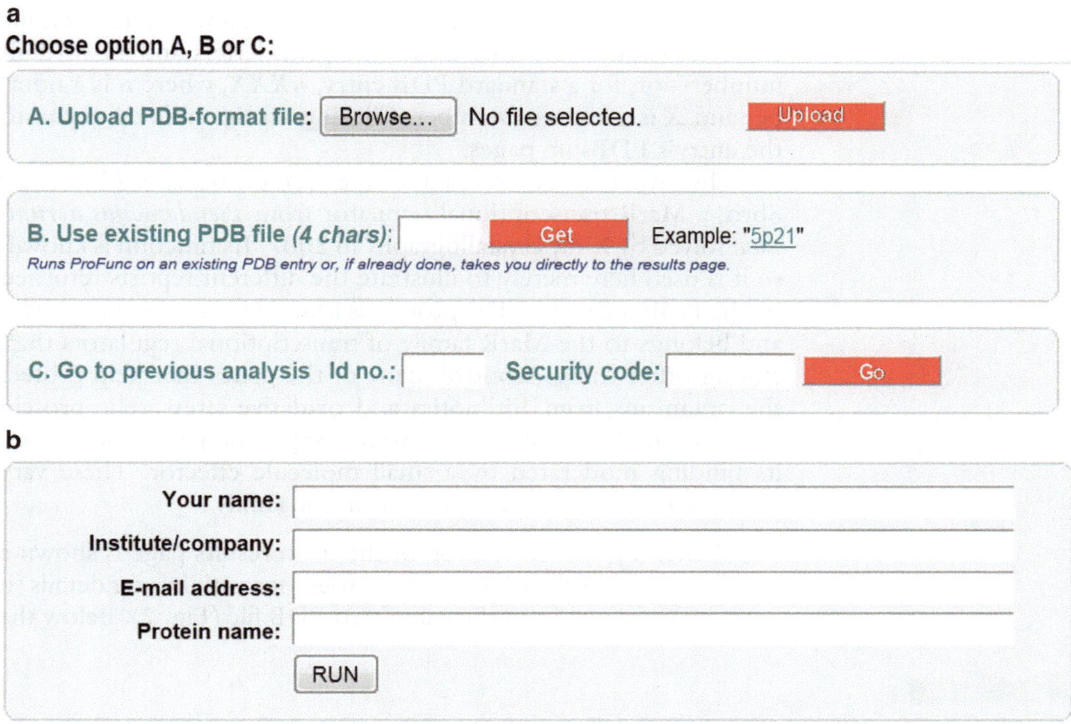

Fig. 1 ProFunc's two entry forms. (**a**) Initial form on ProFunc home page with its three options labeled *A*, *B*, and *C*. Use *A* to upload your own PDB-format file for processing by ProFunc, *B* to run ProFunc on an existing PDB entry (or retrieve the results if the run has already been done), and *C* to go to a completed analysis—for which you need a ProFunc id and security code. (**b**) Options *A* and *B* above take you to this form for entry of your details, including the email address to which the link to the results page and security code will be sent

the protein's name, this will be shown in this box. Otherwise, you can enter its name here (*see* **Note 2**).

2. The **B** option in Fig. 1a accepts the four-character PDB code of an existing PDB entry. If ProFunc has already been run on this entry you will be taken straight to the results page, otherwise the system will ask for your email address and mail you a link when the processing is complete.
3. Finally, the **C** option is for going to the results of a previous ProFunc run. Enter the four-character identifier and security code (as sent you by email) and press **Go** (*see* **Note 3**).

Also on the home page are links to the ProFunc documentation, a tutorial, and example results for PDB entry 5p21 (*see* **Note 4**).

3.2 ProFunc Results

The main results page summarizes the output from each of the sequence- and structure-based searches made, showing the top five hits (where relevant) from each search. Further details can be obtained by clicking on the appropriate icon.

The ProFunc identifier for a user-submitted structure is a four-character code of the form *aa01*—i.e., two letters and a two-digit number—or, for a standard PDB entry, *nXXX,* where *n* is a number and *X* is a letter or number. Clicking on the code takes you to the entry's PDBsum pages.

The protein structure used for the example below is PDB entry 3bro, a MarR transcriptional regulator from *Pseudomonas aeruginosa* solved by X-ray crystallography in 2007. Its function is known, so it is used here merely to illustrate the different reports returned by the ProFunc server. The protein is found in bacteria and archaea and belongs to the MarR family of transcriptional regulators that, among other things, control many of the genes that help protect the organisms from antibiotics and oxidative stress. The protein binds to DNA at specific palindromic or pseudo palindromic sites, its binding moderated by a small molecule effector. These vary widely across the different proteins in the family.

1. *Protein details.* At the top of the main results page is shown a thumbnail image of the protein, together with header details (if any) extracted from the submitted PDB file (Fig. 2). Below the

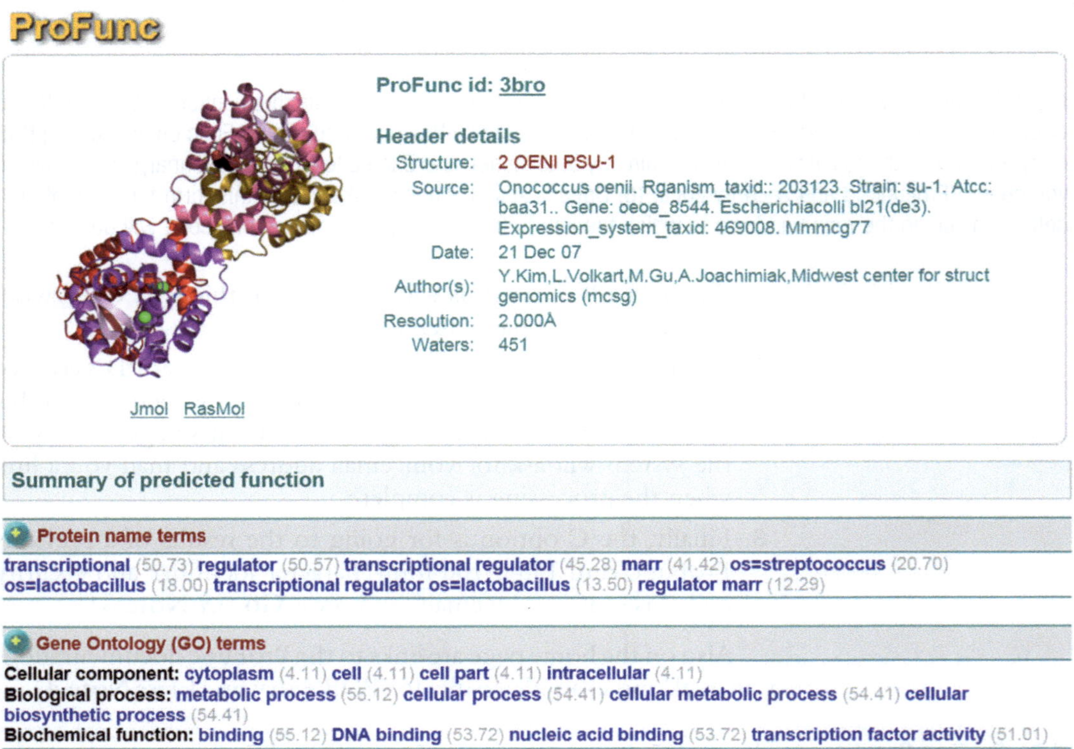

Fig. 2 The *top* part of the ProFunc results page for PDB entry 3bro. The upper half contains a thumbnail image of the structure together with header details taken from the header records in the PDB file. Below that is a summary of the most common terms found in the ProFunc hits from its various searches. The first set comes from the names of the proteins hit, while the second comes from the proteins' Gene Ontology (GO) assignments as given in the corresponding UniProt entry

image are two options for viewing the structure in 3D. The first is a java-based viewer called Jmol, and the other is RasMol. If your browser is correctly configured, clicking either link will display a 3D version of the structure for viewing and manipulating interactively (*see* **Note 5**).

2. *Summary of predicted function.* Below the header details are two sets of terms compiled from all the ProFunc search results (*see* the lower half of Fig. 2). The first group are protein name terms (for example, for 3bro they are "transcriptional regulator," "marr," "os = streptococcus," and so on). The second group are Gene Ontology (GO) terms [28], which are classified by: cellular component, biological process, and biochemical function. The terms are ranked by their significance score which is based on the number of times each term occurs independently in the results and the similarity of the match to the query protein (*see* **Note 6**). Clicking on the "+" icon shows a breakdown of where each term originates from, and its score from the given search method.
3. *List of results.* Below the function summary are listed all the ProFunc methods that were run on the structure and its sequence. Figure 3 shows the first and last of these. Each lists the top five most significant results. Where a score is computed by the method (as in Fig. 3b), the results are ranked by that score and colored according to the score's significance: red = certain match, pink = probable match, orange = possible match, and blue = long shot. For further information about each hit, and for additional hits, click the icon on the left (i.e., the one labeled **InterPro** in Fig. 3a, and **SITE** in Fig. 3b).
4. *Sequence motifs.* The first set of results are from InterProScan and list the various sequence motifs and patterns found in the query protein (Fig. 4). The "wiring diagram" at the top is a schematic plot of the secondary structure of 3bro, with the red jagged elements representing the protein's helices, the yellow arrows its beta strands, and the purple lines its coil regions. The bars below show the extent of each region matched by the InterProScan motifs, and are colored by motif type. The label on the left is the motif identifier or reference code.

   The results are also listed in the table below the diagram. This again shows the reference code and the residue range matched. Clicking the code, or the colored ball in the first column, will take you to the database from which the motif comes (e.g., Pfam, SUPERFAMILY, etc.) where you can find further information about the motif and any functional information associated with it. If you have your browser configured for RasMol (*see* **Note 5**), the icons in the "View in RasMol" column will display the 3D structure in RasMol in backbone

a

Sequence motifs

InterPro

InterPro scan for sequence motifs. Chains A, B, C, D

5 motifs matched in scan against PROSITE, PRINTS, PFam-A, TIGRFAM, PROFILES and PRODOM motifs

	Type	Motif	Name
1.	SUPERFAMILY	SSF46785	
2.	Gene3D	G3DSA:1.10.10.10	
3.	SMART	SM00347	helix_turn_helix multiple antibiotic resistance protein
4.	ProSiteProfiles	PS50995	MarR-type HTH domain profile.
5.	Prints	PR00598x3	Bacterial regulatory protein MarR family signature

b

SITE

Reverse template comparison *vs* structures in PDB.

20 significant hits out of 347 auto-generated templates.

	Score	Template	PDB	Name
1.	780.000	TMP00173	3bro	Crystal structure of the transcription regulator marr from o oeni psu-1
2.	411.312	TMP00056	4ld5	Crystal structure of mepr q18p mutant from multidrug resista aureus clinical isolate
3.	408.312	TMP00056	4l9n	Crystal structure of mepr a103v mutant from multidrug resist aureus clinical isolate
4.	351.188	TMP00056	4l9t	Crystal structure of mepr f27l mutant from multidrug resista aureus clinical isolate
5.	318.969	TMP00156	2fbh	The crystal structure of transcriptional regulator pa3341

... *plus others*

Fig. 3 Two examples of the summary information returned by ProFunc's searches. (**a**) The top five **InterProScan** hits, showing the type, id and name of the matched motif. Clicking the **InterPro** icon on the left takes you to a page showing more detailed information about these matches, as shown in Fig. 4. (**b**) The top five hits from the reverse template search, showing the match score, template identifier, matched PDB entry and its title. For further information, and more hits, click on the **SITE** icon on the *left* to get a full listing as shown in Fig. 6

representation: with thick colored sticks representing the region of the motif, and thin white sticks elsewhere.

For the 3bro example, the motifs picked up include the DNA-binding helix-turn-helix motif (HTH) of the MarR transcription regulator family (Fig. 4). Also picked up is a signature associated with multiple antibiotic resistance.

5. *Matches to existing PDB structures*. The next block of results come from a FASTA search of the query protein's sequence against the proteins in the PDB. The output is a multiple sequence alignment of the top-scoring hits and can tell you if there are structures of similar proteins already in the PDB—which may have functional annotations associated with them.

The default alignment is colored by residue type (blue positive, red negative, etc., as shown in a key at the bottom of the alignment). This coloring can be changed using the drop-down "Annotate by" menu. The options include coloring the sequences by secondary structure, contacts to ligands, residue

a

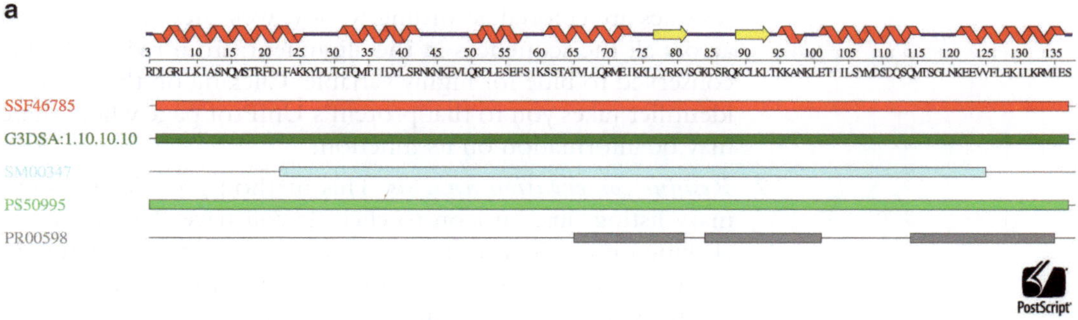

b

Hit	Scan	Reference code	Residue range	View in RasMol	Motif name
1.	SUPERFAMILY	SSF46785	4-136		
2.	Gene3D	G3DSA:1.10.10.10	4-136		
3.	SMART	SM00347	22-124		helix_turn_helix multiple antibiotic resistance protein
4.	ProSiteProfiles	PS50995	3-136		MarR-type HTH domain profile.
5.	FPrintScan	PR00598	84-100 114-134 65-80		Bacterial regulatory protein MarR family signature

Fig. 4 Detailed results from the **InterProScan** search on PDB entry 3bro. (**a**) The schematic "wiring diagram" shows the protein's secondary structure assignments—the red saw tooth being alpha-helical regions and the *yellow arrows* beta-strands. The *colored bars* below show the extent of each domain or motif identified by **InterProScan**, with their identifiers shown on the *left*. (**b**) Table showing the details of each motif, the coloring corresponding to that in the diagram in (**a**). The table gives the motif type, identifier, residue range, and motif name. Clicking on either the colored ball at the left of the table, or on the motif identifier, will take you to the relevant motif description in the given motif type's web site. The small RasMol logos in the table will launch RasMol and show how the domain maps onto the query protein structure. *See* **Note 5** for a comment about viewing in RasMol

similarity, and so on. Press the **APPLY** button to activate your choice (*see* **Note 7**).

It is also possible to change the alignment so that the PDB numbering of the residue positions is shown instead of the default sequence numbering (*see* **Note 8**). You can also opt to include the secondary structure assignments (with helices indicated by a string of "H"s, strands by "E"s, turns by "t"s, and random coil by hyphens), which may show you how well the secondary structures align.

The RasMol icon at the top of the page allows you to select and view the structures from the alignment in RasMol (if your browser allows it), superposed according to the alignment. If the structures superpose well, it is possible that they perform the same, or similar, biological functions.

6. *Sequence search against UniProt*. The other sequence search performed by ProFunc is a BLAST search of the query protein sequence against all the proteins in the UniProt Knowledgebase. An alignment of the top ten matches is shown, followed by a table giving the alignment statistics for the top 50 hits. The

residues are colored according to how well they are conserved across all the sequences in the alignment: from red for highly conserved to blue for highly variable. Clicking on the UniProt identifier takes you to that protein's UniProt page where there may be information on its function.

7. *Residue conservation analysis.* This method provides no summary listing, just an icon to click. If you have RasMol setup, clicking on this icon will show your query protein structure in RasMol colored according to the residue conservation colors in the BLAST alignment from method 6, above.

8. *Matching folds detected by PDBeFold.* This is the first of the structure-based methods that can often provide crucial functional information if the sequence-based methods above have drawn a blank. Structure is more strongly conserved over evolutionary time than sequence [29], and structures having the same fold often perform the same, or similar, functions [30, 31]. Exceptions to this rule include the very common folds, or "superfolds" [32], such as the TIM barrel that is adopted by a wide variety of different proteins that catalyze very many different enzyme reactions.

   The results of the fold match are listed in a table (upper part of Fig. 5), together with match scores and statistics, and are shown schematically in a "wiring diagram" that illustrates the secondary structure similarities (lower part of Fig. 5) (*see* **Note 9**). For every hit, the PDBeFold program performs a 3D superposition on C^{α} atoms of the query structure and the matched structure and calculates the root mean square deviation (r.m.s.d) between the two. The hits are ranked according to *z*-scores computed from the r.m.s.d. values.

   The fold hits for the 3bro structure, shown in Fig. 5, align very well as they all have similar folds that superpose well. However, there are subtle differences between the structures, and it is often these differences that are of most interest as they can help explain differences in substrate specificity. As the MarR proteins are dimeric in nature, small changes in the arrangement of helices in the monomer can have significant implications for the arrangement of the dimer and consequently affect the protein's recognition of its palindromic DNA sequence.

9. *Nest analysis.* Nests are structural motifs that are often found in functionally important regions of protein structures and may be useful for identifying possible functional sites if all the other methods have failed to provide such information. Each has a "score" indicating how functionally significant it is likely to be. The score takes into account the number of NH atoms accessible to solvent, the conservation score of the parent residues, and whether they are associated with one of the larger surface clefts. A score over 2.0 suggests functional significance.

a

Hit	Select	Q-score	Z-score	No. SSE	RMSD (Å)	Seq. id	PDB entry	Name
1	☑	0.691	9.1	8	1.84	27.4%	4l9jA	Crystal structure of s. Aureus mepr in DNA-binding conformat
2	☐	0.651	9.3	8	1.83	27.4%	4lllA	Crystal structure of s. Aureus mepr-DNA complex
3	☐	0.649	9.3	8	1.92	27.4%	4llnA	Crystal structure of s. Aureus mepr-DNA complex
4	☐	0.622	9.0	8	1.85	25.9%	2bv6A	Crystal structure of mgra, a global regulator and major virulence determinant in staphylococcus aureus
5	☐	0.595	9.4	8	1.88	18.5%	1lj9A	The crystal structure of the transcriptional regulator slya
6	☐	0.586	8.3	8	2.20	22.2%	3q5fA	Crystal structure of the salmonella transcriptional regulato complex with DNA
7	☐	0.574	8.3	8	2.40	21.5%	4aijA	Crystal structure of rova from yersinia in complex with a ro promoter fragment
8	☐	0.563	8.2	8	2.46	21.5%	4aikA	Crystal structure of rova from yersinia in complex with an invasin promoter fragment
9	☐	0.550	8.2	8	2.24	17.0%	3s2wA	The crystal structure of a marr transcriptional regulator fr methanosarcina mazei go1
10	☐	0.549	8.6	8	2.14	18.5%	4mnuA	Crystal structure of uncharacterized slya-like transcription from listeria monocytogenes

b

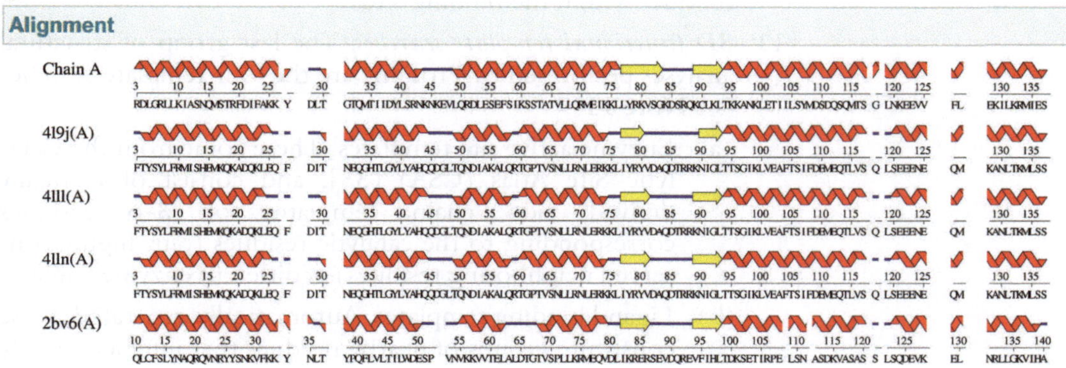

Fig. 5 Fold matches for 3bro returned by PDBeFold. (**a**) The table lists the top ten matches to structures in the PDB, listed in order of *Q*-score—the program's own measure of the significance of each hit. Also shown are a *z*-score, number of matching secondary structure elements (No. SSE), r.m.s.d. for equivalent C^{α} atoms, sequence identity, and name of the matched protein. The checkboxes allow you to select which structures you would like to view in RasMol, superposed on the 3bro structure. (**b**) Schematic diagram showing the match of the secondary structure elements of the structures to that of 3bro

Only two nests are found in the 3bro structure: Lys74-Leu77 and Phe57-Ile59. The latter is the more interesting as it lies between two helices that form a HTH-motif and hence performs an important structural role as part of the motif (*see* **Note 10**).

10. *Cleft analysis.* Protein binding sites are often located in the largest cleft on the protein's surface [23], and this cleft tends to be a good deal larger than the others. Thus, identification of surface clefts can often help identify the protein's functional site. ProFunc uses the SURFNET algorithm to detect gap regions within the protein by fitting spheres of a certain range of sizes between the protein atoms. Once clefts and cavities have been identified they are listed in order of volume, with additional information such as: number of each residue type lining the cleft, numbers of residues at different conservation levels, and any ligands bound inside the cleft. The grayed out columns—showing the number of accessible vertices, numbers

of buried vertices, and average depth of cleft—are not particularly helpful. It is better to display the clefts in 3D using RasMol by selecting the clefts of interest with the checkboxes and clicking **DISPLAY**.

The display options determine how the gap regions are displayed. The **Binding site(s)** option shows each cleft as a closed wireframe, embedded in the protein's surface. The **Binding surface(s)** option shows the protein surface that forms the cleft. With this option the surface can be colored by: gap volume order (i.e., red for largest, purple for second largest, etc.), closest atom type, residue type, residue conservation. The last of these is particularly useful for identifying key residues within the binding site.

11. *3D functional template searches.* The last group of structural analyses performed by ProFunc are the four template searches (*see* **Note 11**):

    (a) Enzyme active site templates. These come from the Catalytic Site Atlas (CSA) [33], and consist of manually derived side chain templates of 3–6 residues corresponding to the catalytic residues (plus highly conserved neighboring residues) of different enzyme families.

    (b) Ligand binding templates. Automatically generated, these constitute groups of nearby residues that interact directly (either via hydrogen bonds or nonbonded contacts) with a bound ligand. Several templates can come from the same structure—particularly where the ligand molecule is a large one—although no two templates can have more than one residue in common. A non-homologous set of protein structures is used when compiling the template database.

    (c) DNA-binding templates. Generated like the ligand binding templates above, but for residues interacting directly with a bound DNA or RNA molecule.

    (d) "Reverse" templates. These are templates automatically generated from the query structure itself, with several hundred three-residue templates typically produced. Residue conservation scores are used to prioritize which residues are selected for each template. The templates are then scanned against a representative dataset of structures in the PDB.

As the 3D templates consist of, typically, just three residues, each search can produce hundreds, or even thousands of hits—particularly in the case of the reverse templates that scan a large chunk of the PDB. So to reduce this list, the hits are scored for relevance and then ranked according to this score. The similarity score is obtained by comparing the environments of the matched template residues in the two proteins—i.e., the protein from which the template came and the one in which the hit was found. After superposing the two structures on the matched residues, all residues within a sphere of 10 Å around the template center are compared and paired if they are of identical or similar residue types. Then the paired residues are filtered to leave only pairs that could have come from a sequence alignment of the two proteins (i.e., the residues in all the pairs are in the same relative sequential order in their respective sequences). Sometimes, many different sets of pairings are possible, so each is scored and the highest scoring one chosen. The score takes into account the number of paired residues in the superposed sequence alignment and the number of insertions that would be required in one or both of the sequences to arrive at this alignment. This proves a very sensitive means of discriminating between true and false positives (*see* **Note 12**).

On its own, the score is not particularly meaningful, so the program computes an *E*-value from the distribution of scores it gets from all the hits in the search. The top hits are then categorized according to the computed *E*-value: certain matches (*E*-value $<10^{-6}$), probable matches ($10^{-6} < E < 0.01$), possible matches ($0.01 < E < 0.10$), and long shots ($0.10 < E < 10.0$).

Figure 6 lists the top hits from the "reverse" template hits for 3bro. The top match here is a self-hit, so not particularly useful. The second hit, to PDB entry 4ld5, is more interesting. It is a "certain match," with *E*-value 8.99×10^{-7}, yet the sequence identity to 3bro is only 27.4%. The red "×19" after the PDB code 4ld5 indicates that 19 of the 3D templates generated from 3bro gave significant matches to 4ld5. Only the scores for the top one of these is shown here. The "Neighbors" column indicates that this had 19 identical residues in the 10 Å region around the template and ten similar ones.

The "Longest fitted segment" column gives the longest span of the superposition of equivalent C^{α} atoms with an r.m.s.d. below 3 Å (the red number gives the number of equivalent residues and the orange number gives the length of the span—i.e., including any gaps in the alignment) (*see* **Note 13**).

The majority of the hits in the table are transcription regulators, and several are associated with multidrug resistance—which is in agreement with what we know for MarR.

For more details on each hit, click either the hit number in the first column, or the colored sphere to the left of it. You will get a separate page, relevant extracts from which are shown in Fig. 7. The statistics shown in the top half (Fig. 7a) are largely the same as on the previous page, but here is also shown the local sequence identity—i.e., the percentage of paired identical

a. Certain matches (E-value < 1.00E-06)

Hit no.	E-value	Similarity score	Neighbours ident/simil [equiv]	Template id	Matched PDB entry	Longest fitted segment	Seq lengths query/target	Overlap	%-tage seq id	Structural similarity
1	0.00E+00	780.00	40/0 [40]	TMP00173	3bro×42: Crystal structure of the transcription regulator marr from o oeni psu-1	134/134	135/135	135	100.00%	99.3%
2	8.99E-07	411.31	19/10 [28]	TMP00056	4ld5×19: Crystal structure of mepr q18p mutant from multidrug resista aureus clinical isolate	100/100	135/138	137	27.41%	99.3%

b. Probable matches (1.00E-06 < E-value < 0.01)

Hit no.	E-value	Similarity score	Neighbours ident/simil [equiv]	Template id	Matched PDB entry	Longest fitted segment	Seq lengths query/target	Overlap	%-tage seq id	Structural similarity
3	1.09E-06	408.31	18/10 [28]	TMP00056	4l9n×17: Crystal structure of mepr a103v mutant from multidrug resist aureus clinical isolate	113/113	135/137	136	26.67%	99.3%
4	4.23E-05	351.19	15/9 [23]	TMP00056	4l9t×17: Crystal structure of mepr f27l mutant from multidrug resista aureus clinical isolate	96/96	135/140	138	27.41%	98.5%
5	6.72E-05	318.97	12/11 [22]	TMP00156	2fbh×5: The crystal structure of transcriptional regulator pa3341	132/132	135/137	138	23.70%	99.2%
6	2.00E-04	253.59	11/8 [17]	TMP00189	3jw4×12: The structure of a putative marr family transcriptional regu clostridium acetobutylicum	79/79	135/123	137	25.20%	99.2%
7	2.06E-04	344.22	12/13 [23]	TMP00243	1s3j×11: X-ray crystal structure of yuso protein from bacillus subtilis	89/89	135/143	145	20.74%	99.2%
8	3.23E-04	337.38	13/10 [23]	TMP00243	3qpt×15: Crystal structure of the salmonella transcriptional regulato	108/108	135/132	137	22.73%	99.2%

Fig. 6 Top eight hits from the reverse template search. (**a**) The first two hits are "certain matches," having an *E*-value $<10^{-6}$. The very first is a self-hit to 3bro, and so can be ignored. The information shown in the table includes the *E*-value, similarity score, numbers of neighboring residues in the vicinity of the template that are of identical or similar residue types in the two structures, PDB code and title of the matched structure (the *red* "$\times N$" gives the total number, *N*, of templates that gave significant matches), the longest segment of the two structures that superpose on C^α atoms with an r.m.s.d. of less than 3 Å, the lengths of the two sequences, their overlap in the alignment, sequence identity and measure of structural similarity (based on the total span of segments in the alignment that can be fitted with an r.m.s.d of less than 3 Å on C^α atoms). Clicking on the small red sphere, or the hit number next to it, takes you to a detailed page for the given his, as shown for Hit 2 in Fig. 7. (**b**) As above, but for the "probable matches," with *E*-values in the range $10^{-6} < E < 0.01$

residues within the 10 Å region around the template. In this case it is 52.8%, which is substantially higher than the overall sequence identity between the two proteins (27.4%). The high degree of similarity between the matched local regions suggests these have been better conserved over evolutionary time than the rest of the proteins and so are likely to be functionally important.

At the bottom of the page (Fig. 7b) is the sequence alignment, and corresponding secondary structures, obtained from the template superposition. Here, the alignment is a particularly good one, with a close correspondence of the secondary structures. The red arrow indicates the longest span of the alignment giving a 3D superposition of the C^α atoms with an r.m.s.d. below 3 Å. The RasMol button below the alignment shows the matched segments in 3D.

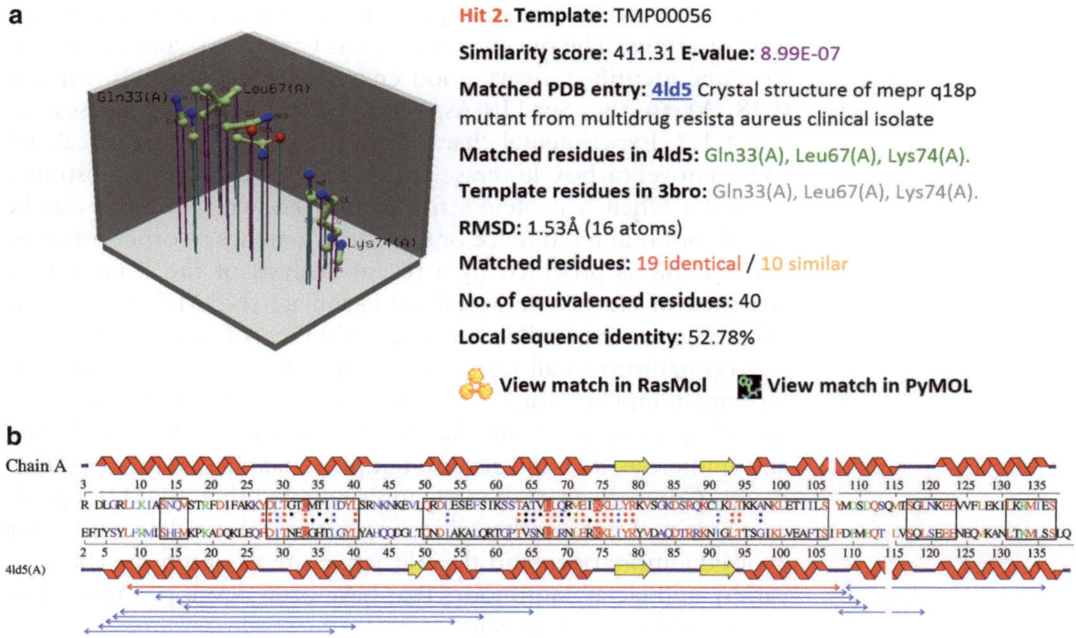

Fig. 7 The detailed information for Hit 2 to PDB entry 4ld5. (**a**) The template residues from 3bro are shown together with the corresponding residues matched in 4ld5, and the r.m.s.d. between their respective atoms. The *red* and *orange* numbers of "Matched residues" correspond to how many equivalent residues in the 10 Å sphere around the template were of identical and similar residue types. Of particular interest is the "Local sequence identity" within this sphere that, at 52.78%, is considerably higher than the overall sequence identity between the two proteins (27.41%). This suggests the template has matched a region of the two proteins that has undergone less change over evolutionary time than the rest of the two proteins, presumably because it is functionally important. High-scoring template matches such as this tend to pick up functional sites because of their importance and the evolutionary constraints of maintaining their key residues. (**b**) A sequence alignment based on the 3D superposition of the structures based on the template match, and identification of equivalent residues. The three matched template residues are highlighted in red, while the equivalenced residues in the surrounding 10 Å sphere are indicated by the *dots* along the center of the alignment (*two dots* for identical residues and one for similar residues). Cases where only one of the sequences has a dot are those where the residues equivalenced in 3D have become separated during the alignment. The *boxed* regions represent segments where the sequence identity of the two sequences exceeds 35%; that is, regions of reasonably significant sequence similarity. The *blue* and *red line* segments identify the structurally "fittable" regions—i.e., where segments from the two proteins can be fitted on their C^{α} atoms with an r.m.s.d. less than 3 Å. The *red* region is the longest such segment (and is only shown if it is at least 20 residues in length)

4 Case Studies

One of the first applications of ProFunc was in helping determine the function of BioH from *Escherichia coli* [34], a protein known to be involved in biotin synthesis. Its 3D structure was solved by X-ray crystallography, but its biochemical function was unknown. The structure consisted of a Rossmann fold—one of the "superfolds" found in numerous enzymes performing a variety of different catalytic reactions. The question was: what type of enzyme was BioH? ProFunc identified a very good enzyme template match (r.m.s.d. 0.28 Å) to the Ser-His-Asp catalytic triad of the lipases, E. C.3.1.1.3. Experimental characterization of the protein revealed it to be a novel carboxylesterase acting on short acyl chain substrates.

Even when a protein's function is known, ProFunc can be useful for finding sequence or structural matches to other proteins. An example is provided by a recent analysis of the X-ray crystal structure of New Delhi Metallo-β-lactamase (NDM-1), PDB code 3rkj, a gene responsible for the resistance of many pathogenic microorganisms to all known β-lactam antibiotics [35]. Here, the enzyme templates found a clear match to the active site of a metallo-β-lactamase from *Bacteroides fragilis* (PDB code 2bmi), and numerous fold and reverse template matches to other β-lactamases. Comparison of the structures revealed that NDM-1's proficiency at nullifying such a wide range of antibiotics lay in it having evolved a more open and flexible active site to deactivate the wide variety of β-lactam antibiotics that have been developed to combat pathogenic microorganisms.

5 Notes

1. *PDB format.* This format was originally introduced when the PDB was established in the 1970s [36]. It is a simple text-based format with fixed-width columns (originally formulated for use with punched cards). Its definition can be found at:

   http://www.wwpdb.org/documentation/file-format.

   It has recently been superseded by mmCIF [37] to deal with its shortcomings (e.g., field sizes too small to cope with the very large protein complexes that are now regularly solved). However, a lot of legacy software still uses the old format, and it is the format accepted by ProFunc.

   There are a number of common errors that users' files occasionally contain. (This is usually the fault of the software that generated the files—e.g., homology modeling, docking, and molecular dynamics programs. It is always surprising that such a simple format should be beyond the capability of many of these programs).

   One such error is the failure to assign unique atom names to ligand atoms. Thus all carbon atoms are named "C," all oxygen atoms "O," and so on. PDB format requires unique names, such as "C1," "C2," "C3," and so on.

   Another error is that the file contains two (or more) protein chains superposed on one another. This often occurs when a homology modeling program writes out the template structure as well as the model it has generated from it. Having two interpenetrating structures in the file confuses the programs. Protein chains with identical or blank chain identifiers also confuse the programs as this goes against the PDB format rules.

There are others. In short, if the output you get from ProFunc and/or PDBsum looks rather odd or incomplete in some respect, it might be due to an incorrectly formatted input file.

2. *Protein name.* It is always useful to have your protein's name in the PDB file you are submitting, particularly if you are submitting more than one structure. It is helpful in identifying which set of results belong to which structure. The standard way to do this is to have a set of header records at the start of the file. Lines beginning with HEADER, TITLE, and COMPND contain information about the protein and will appear on the results pages in both ProFunc and PDBsum. You can refer to the PDB file format (*see* **Note 1**) above, or download any entry from the PDB to see how these lines are used.
3. *Retention of the results.* The ProFunc and PDBsum results are only stored for a limited period of time (currently about 3 months) before being deleted. This is because the data take up a lot of disk space that needs to be freed every now and then. Furthermore, the data are stored on a filer that is not backed up at all so, in principle, could vanish at any time without the possibility of recovery. This has not yet occurred, but if your data are lost, you can always resubmit your structure and have the results generated anew.
4. *ProFunc documentation and tutorial.* At the time of writing the documentation was a little out of date in terms of certain analyses no longer being part of the ProFunc suite of programs. In particular, the Tempura server, mentioned in the tutorial, no longer exists.
5. *3D viewers.* Because of concerns over security issues, modern-day web browsers are a lot stricter about what can and cannot be done than they used to be. Consequently, it can be quite difficult to configure your browser to use either Jmol or RasMol. The former is a Java application and, because of security fears about Java, some institutes completely block all Java programs from their browsers. If using your own personal computer you can find how to allow Java content on your browser by searching Google for, say, "enabling java in firefox." If you have Java installed, Jmol should then work.

As for running RasMol from a browser, this used to be a fairly straightforward procedure involving creation of a new MIME type in your browser and associating it with a batch script to call RasMol. It is no longer possible to create new MIME types, but you can try the instructions at:

http://www.ebi.ac.uk/pdbsum/none&template=doc_viewers.html.

Failing that, if you save the file you get when you click on the RasMol link (call it something like *rasmol.scr*), you can load RasMol separately and type "**script rasmol.scr**" into its command window (adding the appropriate path to wherever the file is stored).

It is planned to eventually add new viewer options to get around these security issues.

6. *Summary terms.* This ProFunc feature was requested by a number of early users, but I must admit I have never found it particularly useful.
7. *Coloring of PDB sequence alignment.* The most useful coloring scheme for the alignment is probably Residue Similarity (which ranges from red for most similar to black for least) as this indicates the regions of the matched proteins that are the best match to the query protein. The statistics in the table at the end of the page show the match scores in terms of sequence identity, *z*-score, and *E*-value. The last of these is particularly helpful as extremely low values can indicate that even sequences that have apparently low sequence identity may still constitute a significant match. The first few hits for 3bro typify this. Usually, a sequence identity of >40% is indicative of a significant match, but here the sequence identities are around 28% yet have *E*-values of around 10^{-7}, which indicates the matches are significant.
8. *Residue numbering.* The residues in a PDB file often appear to be numbered according to some eccentric whim of the depositors. They can be negative, zero, skip one or more, or even have letter suffixes (e.g., 30A, 30B). It is not always easy to match a given residue in the alignment to the corresponding residue in the PDB file unless one switches on the option to number each sequence separately.
9. *Fold-match wiring diagram.* For very long sequences the wiring diagram may be rather compressed and difficult to see clearly. You can click the PostScript icon below it to view the original PostScript image and zoom in for a better view. The diagram shows unmatched secondary structure elements (SSEs) in white and, occasionally, it may appear that PDBeFold has missed some obvious SSE matches, but this is simply because the algorithms for calculating secondary structures differ between PDBeFold and ProFunc.
10. *Nests with functional significance.* For an interesting nest match you can look at the ProFunc results for PDB entry 5p21. Here, the most significant nest belongs to a large loop binding the ligand and is a functionally important motif known as a P-loop, which is used by many proteins to bind ATP.

11. *3D template searches.* Of all ProFunc's methods the template searches are often the most informative and useful as they can pick up functionally important sites and identify the key residues involved.

12. *Template scoring method.* The local similarity score used to rank the hits from the 3D template searches is capable of identifying even quite distant homologues having a very low overall sequence similarity. This happens when the search picks up a functional site that has been well conserved over evolutionary time despite the rest of the proteins having markedly diverged (in sequence, and possibly also structure). So, for example, proteins having a sequence identity of 20–30% overall can have a much higher sequence identity, of say 40–50%, in the region of the functional site.

13. *Longest fitted segment.* This measure can sometimes highlight hits that do not look promising according to the other match scores but which may be worth looking at. So, a hit that is a "long shot" according to the *E*-value may result in a good structural fit and hence show a long segment of the two proteins that superpose on their C^{α} atoms with an r.m.s.d. of less than 3 Å.

Acknowledgments

I would like to thank James Watson for writing the ProFunc tutorial on which parts of this chapter are based.

References

1. Grabowski M, Niedzialkowska E, Zimmerman MD, Minor W (2016) The impact of structural genomics: the first quindecennial. J Struct Funct Genom 17(1):1–16. doi:10.1007/s10969-016-9201-5
2. Lee D, de Beer TA, Laskowski RA, Thornton JM, Orengo CA (2011) 1,000 structures and more from the MCSG. BMC Struct Biol 11:2. doi:10.1186/1472-6807-11-2
3. Jaroszewski L, Rychlewski L, Li Z, Li W, Godzik A (2005) FFAS03: a server for profile—profile sequence alignments. Nucleic Acids Res 33(Web Server issue):W284–W288. doi:10.1093/nar/gki418
4. Binkowski TA, Freeman P, Liang J (2004) pvSOAR: detecting similar surface patterns of pocket and void surfaces of amino acid residues on proteins. Nucleic Acids Res 32(Web Server issue):W555–W558. doi:10.1093/nar/gkh390
5. Pal D, Eisenberg D (2005) Inference of protein function from protein structure. Structure 13(1):121–130. doi:10.1016/j.str.2004.10.015
6. Laskowski RA, Watson JD, Thornton JM (2005) ProFunc: a server for predicting protein function from 3D structure. Nucleic Acids Res 33(Web Server issue):W89–W93
7. Laskowski RA, Hutchinson EG, Michie AD, Wallace AC, Jones ML, Thornton JM (1997) PDBsum: a web-based database of summaries and analyses of all PDB structures. Trends Biochem Sci 22(12):488–490. doi:10.1016/S0968-0004(97)01140-7
8. de Beer TA, Berka K, Thornton JM, Laskowski RA (2014) PDBsum additions. Nucleic Acids Res 42(Database issue):D292–D296. doi:10.1093/nar/gkt940
9. Pearson WR (1990) Rapid and sensitive sequence comparison with FASTP and FASTA. Methods Enzymol 183:63–98
10. Berman HM, Kleywegt GJ, Nakamura H, Markley JL (2014) The Protein Data Bank

archive as an open data resource. J Comput Aided Mol Des 28(10):1009–1014. doi:10.1007/s10822-014-9770-y

11. Altschul SF, Madden TL, Schaffer AA, Zhang J, Zhang Z, Miller W, Lipman DJ (1997) Gapped BLAST and PSI-BLAST: a new generation of protein database search programs. Nucleic Acids Res 25(17):3389–3402
12. UniProt C (2015) UniProt: a hub for protein information. Nucleic Acids Res 43(Database issue):D204–D212. doi:10.1093/nar/gku989
13. Valdar WS, Thornton JM (2001) Conservation helps to identify biologically relevant crystal contacts. J Mol Biol 313(2):399–416. doi:10.1006/jmbi.2001.5034
14. Mitchell A, Chang HY, Daugherty L, Fraser M, Hunter S, Lopez R, McAnulla C, McMenamin C, Nuka G, Pesseat S, Sangrador-Vegas A, Scheremetjew M, Rato C, Yong SY, Bateman A, Punta M, Attwood TK, Sigrist CJ, Redaschi N, Rivoire C, Xenarios I, Kahn D, Guyot D, Bork P, Letunic I, Gough J, Oates M, Haft D, Huang H, Natale DA, Wu CH, Orengo C, Sillitoe I, Mi H, Thomas PD, Finn RD (2015) The InterPro protein families database: the classification resource after 15 years. Nucleic Acids Res 43(Database issue): D213–D221. doi:10.1093/nar/gku1243
15. Sigrist CJ, de Castro E, Cerutti L, Cuche BA, Hulo N, Bridge A, Bougueleret L, Xenarios I (2012) New and continuing developments at PROSITE. Nucleic Acids Res 41(Database issue):D344–D347. doi:10.1093/nar/gks1067
16. Finn RD, Coggill P, Eberhardt RY, Eddy SR, Mistry J, Mitchell AL, Potter SC, Punta M, Qureshi M, Sangrador-Vegas A, Salazar GA, Tate J, Bateman A (2016) The Pfam protein families database: towards a more sustainable future. Nucleic Acids Res 44(D1): D279–D285. doi:10.1093/nar/gkv1344
17. Letunic I, Doerks T, Bork P (2015) SMART: recent updates, new developments and status in 2015. Nucleic Acids Res 43(Database issue): D257–D260. doi:10.1093/nar/gku949
18. Attwood TK, Coletta A, Muirhead G, Pavlopoulou A, Philippou PB, Popov I, Roma-Mateo C, Theodosiou A, Mitchell AL (2012) The PRINTS database: a fine-grained protein sequence annotation and analysis resource—its status in 2012. Database (Oxford) 2012: bas019. doi:10.1093/database/bas019
19. Gough J, Karplus K, Hughey R, Chothia C (2001) Assignment of homology to genome sequences using a library of hidden Markov models that represent all proteins of known structure. J Mol Biol 313(4):903–919. doi:10.1006/jmbi.2001.5080
20. Servant F, Bru C, Carrere S, Courcelle E, Gouzy J, Peyruc D, Kahn D (2002) ProDom: automated clustering of homologous domains. Brief Bioinform 3(3):246–251
21. Krissinel E, Henrick K (2004) Secondary-structure matching (SSM), a new tool for fast protein structure alignment in three dimensions. Acta Crystallogr D Biol Crystallogr 60 (Pt 12 Pt 1):2256–2268. doi:10.1107/S0907444904026460
22. Laskowski RA (1995) SURFNET: a program for visualizing molecular surfaces, cavities, and intermolecular interactions. J Mol Graph 13 (5):323–330 . doi:10.1016/0263-7855(95)00073-9307–328
23. Laskowski RA, Luscombe NM, Swindells MB, Thornton JM (1996) Protein clefts in molecular recognition and function. Protein Sci 5 (12):2438–2452. doi:10.1002/pro.5560051206
24. Watson JD, Milner-White EJ (2002) A novel main-chain anion-binding site in proteins: the nest. A particular combination of phi,psi values in successive residues gives rise to anion-binding sites that occur commonly and are found often at functionally important regions. J Mol Biol 315(2):171–182. doi:10.1006/jmbi.2001.5227
25. Watson JD, Milner-White EJ (2002) The conformations of polypeptide chains where the main-chain parts of successive residues are enantiomeric. Their occurrence in cation and anion-binding regions of proteins. J Mol Biol 315(2):183–191. doi:10.1006/jmbi.2001.5228
26. Laskowski RA, Watson JD, Thornton JM (2005) Protein function prediction using local 3D templates. J Mol Biol 351(3):614–626
27. Barker JA, Thornton JM (2003) An algorithm for constraint-based structural template matching: application to 3D templates with statistical analysis. Bioinformatics 19 (13):1644–1649
28. Ashburner M, Ball CA, Blake JA, Botstein D, Butler H, Cherry JM, Davis AP, Dolinski K, Dwight SS, Eppig JT, Harris MA, Hill DP, Issel-Tarver L, Kasarskis A, Lewis S, Matese JC, Richardson JE, Ringwald M, Rubin GM, Sherlock G (2000) Gene ontology: tool for the unification of biology. The Gene ontology consortium. Nat Genet 25(1):25–29. doi:10.1038/75556
29. Chothia C, Lesk AM (1986) The relation between the divergence of sequence and structure in proteins. EMBO J 5(4):823–826

30. Todd AE, Orengo CA, Thornton JM (1999) Evolution of protein function, from a structural perspective. Curr Opin Chem Biol 3 (5):548–556
31. Todd AE, Orengo CA, Thornton JM (2001) Evolution of function in protein superfamilies, from a structural perspective. J Mol Biol 307 (4):1113–1143
32. Orengo CA, Jones DT, Thornton JM (1994) Protein superfamilies and domain superfolds. Nature 372(6507):631–634. doi:10.1038/372631a0
33. Porter CT, Bartlett GJ, Thornton JM (2004) The catalytic site atlas: a resource of catalytic sites and residues identified in enzymes using structural data. Nucleic Acids Res 32(Database issue):D129–D133
34. Sanishvili R, Yakunin AF, Laskowski RA, Skarina T, Evdokimova E, Doherty-Kirby A, Lajoie GA, Thornton JM, Arrowsmith CH, Savchenko A, Joachimiak A, Edwards AM (2003) Integrating structure, bioinformatics, and enzymology to discover function: BioH, a new carboxylesterase from *Escherichia coli*. J Biol Chem 278(28):26039–26045. doi:10.1074/jbc.M303867200
35. Kim Y, Tesar C, Mire J, Jedrzejczak R, Binkowski A, Babnigg G, Sacchettini J, Joachimiak A (2011) Structure of apo- and monometalated forms of NDM-1—a highly potent carbapenem-hydrolyzing metallo-beta-lactamase. PLoS One 6(9):e24621. doi:10.1371/journal.pone.0024621
36. Bernstein FC, Koetzle TF, Williams GJ, Meyer EF Jr, Brice MD, Rodgers JR, Kennard O, Shimanouchi T, Tasumi M (1977) The Protein Data Bank: a computer-based archival file for macromolecular structures. J Mol Biol 112 (3):535–542
37. Bourne PE, Berman HM, McMahon B, Watenpaugh KD, Westbrook JD, Fitzgerald PM (1997) Macromolecular crystallographic information file. Methods Enzymol 277:571–590

Chapter 8

G-LoSA for Prediction of Protein-Ligand Binding Sites and Structures

Hui Sun Lee and Wonpil Im

Abstract

Recent advances in high-throughput structure determination and computational protein structure prediction have significantly enriched the universe of protein structure. However, there is still a large gap between the number of available protein structures and that of proteins with annotated function in high accuracy. Computational structure-based protein function prediction has emerged to reduce this knowledge gap. The identification of a ligand binding site and its structure is critical to the determination of a protein's molecular function. We present a computational methodology for predicting small molecule ligand binding site and ligand structure using G-LoSA, our protein local structure alignment and similarity measurement tool. All the computational procedures described here can be easily implemented using G-LoSA Toolkit, a package of standalone software programs and preprocessed PDB structure libraries. G-LoSA and G-LoSA Toolkit are freely available to academic users at http://compbio.lehigh.edu/GLoSA. We also illustrate a case study to show the potential of our template-based approach harnessing G-LoSA for protein function prediction.

Keywords Structure-based function prediction, Protein local structure comparison, Structure similarity, Structure library search, G-LoSA toolkit

1 Introduction

A major goal of molecular biology is to annotate the functions of all proteins encoded in a genome to understand life at the molecular level and develop effective pharmaceuticals. In recent years, advanced sequencing techniques have exponentially expanded the number of protein sequences available in databases. However, reliable function annotations have been assigned to only a small fraction of all proteins [1]. UniProtKB/Swiss-Prot and UniProtKB/TrEMBL contain 61,522,041 protein sequence entries as of February 17, 2016, but the number of entries with annotated function at high accuracy is 427,210 (0.69%). Due to their inherent time-, cost-, and labor-intensiveness, experimental methods for protein function annotation are still far from high-throughput

Daisuke Kihara (ed.), *Protein Function Prediction: Methods and Protocols*, Methods in Molecular Biology, vol. 1611,
DOI 10.1007/978-1-4939-7015-5_8,

characterization [2]. Therefore, computational prediction has emerged as an alternative approach.

Sequence-based function prediction is based on characterizing the sequence similarity of a target protein to the protein sequences whose functions were previously identified. However, this approach usually becomes unreliable with lower levels of sequence similarity [3]. In the case that the three-dimensional (3D) structure of a target protein is available, function annotation based on its structure is the method of choice to enhance the prediction accuracy. Indeed, an increasing number of protein structures are available without knowledge of their functions through advanced high-throughput structure determination techniques [4]. In addition to the experimental structure determination, outstanding advances in computational protein structure prediction from protein sequences are also enriching the universe of available structural data [5]. Therefore, it is imperative to develop efficient computational methods for structure-based protein function prediction to reduce the knowledge gap between available protein structure data and their function annotations.

The identification of a small molecule ligand binding site (BS) and ligand structure is a key step for the determination of a protein's molecular function. A predicted ligand BS can be used to assign BS residues in an uncharacterized protein, making it possible to infer the function from characterized proteins by a structural similarity comparison between their BS structures [6]. In addition, the identification of a natural small molecule ligand such as a substrate and/or an effector is a basis for predicting the function of enzymes [7].

A rapid increase in the number of protein-ligand complex structures (structure templates) in the Protein Data Bank (PDB, http://www.rcsb.org) [8] provides insight into small molecule ligand binding that is evolutionarily distant but structurally homologous to a target protein. A conventional approach is to identify structure templates through global structure comparison between a target and template proteins based on the concept that a protein pair with similar global structures has a similar function, and thus they may possess similar ligand BS [9, 10]. However, molecular recognition by small molecule ligands occurs in a local region on the protein surface, not the global structure. Therefore, local structure-focused comparison is an approach toward accurately characterizing the similarity of ligand BS structures between different proteins. Indeed, an all-against-all comparison of SCOP (structural classification of proteins) [11] representatives showed the cases of evolutionary convergence to common functional sites from different folds, indicating that overall protein structure dissimilarities do not necessarily imply dissimilarities in their ligand BS [12].

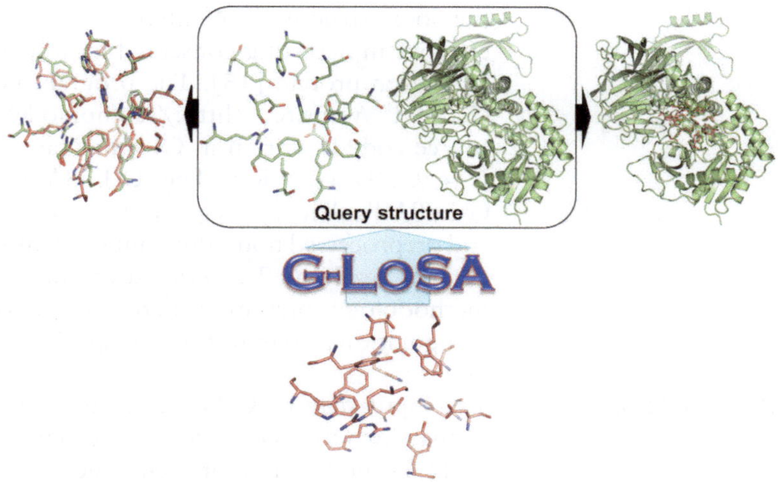

Fig. 1 G-LoSA for local structure alignment and similarity measurement. A local or whole protein structure can be used as a query structure to align another local structure

Here, we present a computational methodology for predicting small molecule ligand BS and ligand structure utilizing G-LoSA (Graph-based Local Structure Alignment) [13–15], our robust large-scale protein local structure alignment and similarity measurement tool (Fig. 1). G-LoSA is used as a search engine for identifying template holo-protein structures that have structurally similar ligand BS to a local region in a query (target) protein. For a given query protein, each BS structure in a structure library is optimally aligned onto the protein surface by G-LoSA, and the structure similarity at the aligned region is measured by GA-score (G-LoSA Alignment score, *see* **Note 1**). After an entire library search, a set of BS templates is then identified in terms of GA-score. The template ligands are mapped onto the query protein upon the structure alignment of the template BS. A ligand BS and a structure of the query protein are predicted based on the mapped template ligands. A biological function of the query protein is interrogated from characterized knowledge of the template proteins.

2 Materials

2.1 G-LoSA and G-LoSA Toolkit

G-LoSA is a computational tool to align protein local structures in a sequence order independent way and to provide a GA-score, a size-independent and chemical feature (CF)-based quantity of structural similarity for a given structure pair (local structure vs. local structure or local structure vs. whole structure). G-LoSA is a robust tool that can be used to accurately quantify structural similarity for a given local structure pair regardless of the size and characteristics of the local structures of interest. In addition, G-LoSA is highly effective in detecting conserved local regions on the entire surface of a given protein [15]. The program G-LoSA is available at the G-LoSA website (http://compbio.lehigh.edu/GLoSA). The source code is written in C++ and can be easily compiled using a C++ GNU compiler. The G-LoSA website also provides the G-LoSA Toolkit, a package of standalone Java application programs and preprocessed nonredundant small molecule ligand and their BS libraries (*see* **Note 2**), so that users may not only easily practice the methodology presented here, but also implement their own computational systems for user-specific research purposes.

2.2 Input Data

The input data of G-LoSA are three protein structures in PDB format (two for query protein and one for library BS structure, *see* **Note 3**) and CF files for each structure (*see* **Note 1**). A CF file has information to describe CFs in each amino acid and their coordinates for a given protein structure. For the two input PDB files for query protein, one is the whole structure of a single protein for which a user wants to predict the ligand/BS structure, and the other is the structure of its surface residues (*see* Subheading 3.1). The library BS structures are from a nonredundant PDB small molecule ligand/BS structure library. Each library consists of PDB files for BS and ligand structures and CF files for the BS structures.

3 Methods

3.1 Prepare Input Query Structures

1. *Generate surface residues file*: Searching for an optimum structural alignment of a library BS onto a whole protein becomes computationally expensive when the size of the protein is big (e.g., >400 amino acids). The alignment space can be reduced by focusing only on solvent-exposed surface regions because they are the real targets in the query protein for the alignment of library BS structures. Therefore, G-LoSA Toolkit provides the *GetSurfaceResiduesUsingVisGrid* program to extract surface residues from a given protein structure. This application utilizes an external program VisGrid (http://kiharalab.org/VisGrid/) that uses the visibility criterion to characterize local geometric features of protein surfaces [16]. In some cases, certain BS residues are missing in the surface residues file due to the incompleteness of surface/pocket residue detection tools, which can result in an inaccurate GA-score calculation. To solve this problem, G-LoSA needs both surface residues and whole protein as input structures for alignment and scoring, respectively.

2. *Generate chemical feature files*: Users can generate CF files for each input structure using the *GenerateCFFiles* program in G-LoSA Toolkit.

3.2 Search PDB Structure Library

For structure alignment and similarity measurement by G-LoSA for a large number of library BS structures, users need to prepare a script to run a batch job. The *GLoSASearch* program in G-LoSA Toolkit enables users to easily carry out this task. To run this program, users need to prepare an input file (an example file is available at the G-LoSA web site) for *GLoSASearch* to read a set of parameters defining below:

- Query protein structure file.
- CF file for the query protein structure file.
- Surface residues file for the query protein structure file.
- CF file for the surface resides file.
- Directory of small molecule ligand/BS structure library.
- List of library BS structures.
- Directory where CF files for library BS structures are saved.
- G-LoSA execution file.
- Cutoff GA-score for writing rotation matrix (0.6 is recommended).

G-LoSA calculations over an entire structure library need long computational time when a large protein is used as the query. In this case, it is recommended to submit multiple sub-batch jobs if a multi-core computer or a cluster computer is available. Users can easily divide the list of library BS structure into a number of sub-lists, submit each batch job for each sub-list, and then merge results from each sub-batch. For this multiple job handling, users can use the G-LoSA Toolkit programs such as *SplitList*, *RunPBSJobs*, and *MergeGLoSASearchResults*. Each sub-batch job should be run in individual directories to avoid overwriting the results.

3.3 Get Templates

G-LoSA calculations against a structure library by *GLoSASearch* provide two files: one is a list of GA-score for each library BS structure and the other is a list of rotation matrix for BS structures whose GA-score is $\geq$ a cutoff.

1. *Get a list of good templates*: To obtain a list of templates with high GA-score, users can use the *GetTopTemplates* program in G-LoSA Toolkit. This application sorts the results in the GA-score list and provides a list of BS templates selected by user-defined GA-score cutoff and number of templates.
2. *Get template structures aligned onto query protein*: For prediction and visual inspection, it is needed to generate template structures aligned onto a query protein by G-LoSA. This can be

achieved by the *GetAlignedTemplates* program. For a given list of templates, this program generates BS and its ligand structures aligned onto the query protein using the given rotation matrix.

3. *Filter templates*: The identified ligand templates could have structural similarity from one another, which could make it inconvenient to do the visual inspection and biological function retrieval. The *FilterTemplatesByPositionalOverlapTanimoto* program in G-LoSA Toolkit removes redundant templates based on structural similarity between their ligands, quantified by the positional overlap Tanimoto coefficient (T_{PO}) (*see* **Note 4**). A value 0.7 is used as the cutoff T_{PO}.
4. *Generate PyMol script for visualization*: For a manual structure comparison, users may want to load all of the identified BS and ligand templates with the query protein on a molecular visualization program. The *GeneratePyMolScript* program in G-LoSA Toolkit generates a PyMol script (scrip_name.pml) to visualize the query and template structures for a given template list.

3.4 Retrieve Biological Functions of Templates

The PDB provides annotation information for each PDB entry, encompassing Domain Annotation based on the SCOP [11] and CATH [17] database, Protein Family Annotation based on the Pfam database [18], and Gene Product Annotation from Gene Ontology Consortium (http://geneontology.org). The *GetAnnotations* program in G-LoSA Toolkit generates a HTML file from the template list file. Users can easily retrieve the annotation information provided by the PDB using the links in the HTML file.

3.5 Case Study

We show a representative example to illustrate the application of the described method using RNA-helicase Prp43 from *Chaetomium thermophilum* (PDB:5D0U) [19] as a query protein (Fig. 2). We selected this protein because the size is big (705 amino acids), the structure has been recently released, and its native ligand is adenosine-5′-diphosphate (ADP), which is a hard target for function prediction due to the broad binding of the ligand to many distinct proteins involved in various functions. We performed a G-LoSA search against nonredundant small molecule ligand/BS structure library, which was prepared using a sequence identity cutoff of 50% and a Tanimoto coefficient of 0.7. For the library search under a strict condition, we excluded all BS structures from homologous proteins using 30% sequence identity to the query protein. We identified a set of templates whose GA-score is $\geq$0.7 and removed redundant templates using a T_{PO} cutoff of 0.7 and finally selected five templates with the highest GA-score. The aligned structures of the top five templates are shown in Fig. 3a–e. All the template BS structures are well aligned onto the native one of the query protein, showing high local structural similarity. In particular, the ligand

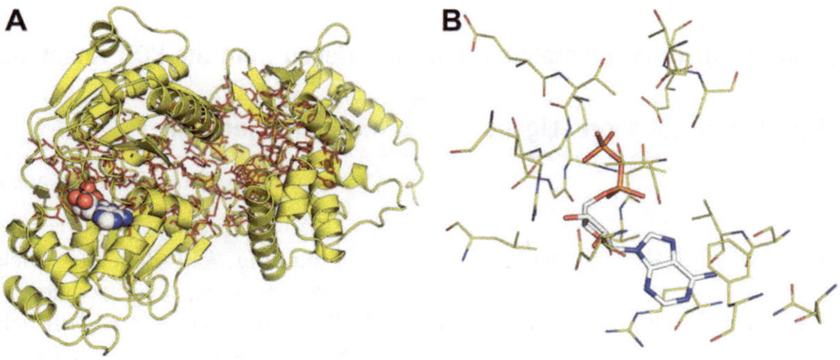

Fig. 2 The structure of query protein used for a case study (PDB:5D0U). (**a**) Overall topology of the protein shown in cartoon representation with the native ligand using sphere representation. The surface residues used for G-LoSA search are also displayed in *red* using stick representation. (**b**) The structures of BS and its native ligand are shown as *yellow lines* and *white sticks*, respectively

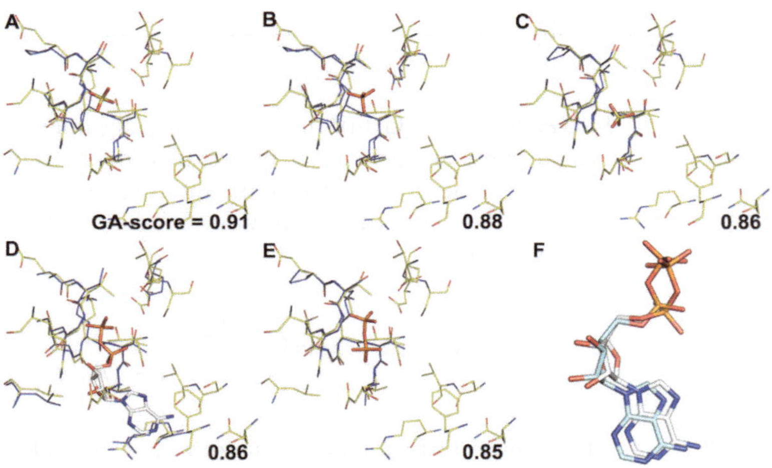

Fig. 3 The aligned structures of the top five templates (ligand and BS) onto the query protein. (**a–e**) The native BS residues of the query protein (*yellow lines*) and each template (*blue lines*) are shown with the template ligands as sticks. (**f**) A comparison of structure and conformation between the native and template ligand. The structure of the template ligand (*cyan*) was obtained from (**d**)

structure of the fourth best template is exactly the same as that of the native ligand (Fig. 3f). Table 1 describes the molecular functions of the query protein and the templates retrieved from the PDB Annotations. The first, second, and fourth templates accurately predict the ATP-dependent helicase function of the query protein, demonstrating the potential of our template-based approach harnessing G-LoSA for protein function prediction.

Table 1
Molecular functions of query and template proteins retrieved from the PDB Annotation

Type	PDB ID	Chain	GA-score	Ligand	Molecular function
Query	5D0U	A		ADP	Nucleotide binding, Helicase activity, ATP binding, Hydrolase activity
Template	3O8B	A	0.91	Sulfate	ATP binding, ATP-dependent helicase activity, Serine-type peptidase activity
	1CU1	A	0.88	Phosphate	ATP binding, ATP-dependent helicase activity, Serine-type peptidase activity
	2IT1	B	0.86	Sulfate	Nucleotide binding, ATP binding, ATPase activity
	3LLM	A	0.86	ADP	Nucleic acid binding, ATP binding, ATP-dependent helicase activity
	1G29	1	0.85	Pyrophosphate	Nucleotide binding, Transport activity, ATP binding, ATPase activity

4 Notes

1. **GA-score**: In G-LoSA, all possible alignments between two structures are generated by iterative maximum clique search and fragment superposition, and the optimal alignment is determined by the maximum GA-score [15]. GA-score is defined by

$$\text{GA-score} = \text{Max}\left[\frac{1}{N_\text{T}}\left(\sum_{i}^{N_\text{ali}}\frac{q_i}{1+(d_i/d_0)^2}\right)\right]$$

$$q_i = \begin{cases} 1 & \text{if chemical feature types are identical, or HD/OH or HA/OH} \\ 0.8 & \text{if HD/PC, HA/NC, OH/PC, OH/NC, or AR/AL} \\ 0.5 & \text{otherwise} \end{cases} \quad (1)$$

where "Max" denotes that the GA-score is the maximum of all possible alignments, N_T is the smaller number of CF points (CFPs) between two local structures, and N_ali is the number of aligned CFPs. d_i is the distance between the CFPs in the ith pair. d_0 is the scaling factor to normalize the aligned distances. q_i is defined based on the chemical feature similarity of the ith CFP pair. We define seven different CFs for amino acids: hydrogen bond donor (HD), hydrogen bond acceptor (HA), hydroxyl group (OH), positively charged atom (PC), and negatively charged atom (NC) are defined by single atom points, while aromatic ring (AR) and aliphatic hydrophobic group (AL) are defined by the geometric center of a set of atoms (Fig. 4).

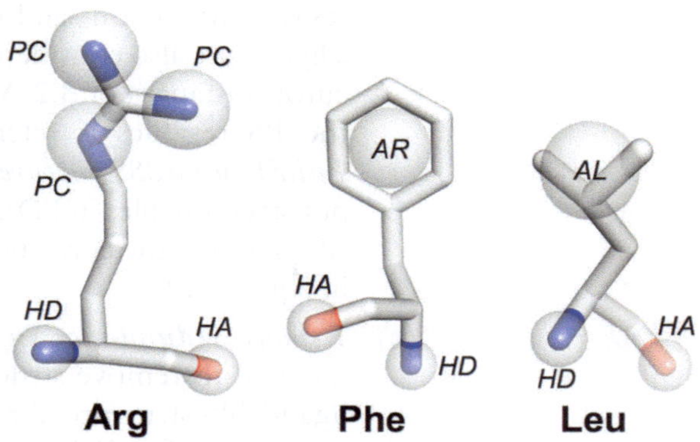

Fig. 4 Illustration of chemical feature points for three representative amino acids (Arg, Phe, and Leu)

2. **Prepare PDB structure library for small molecule ligands and their binding sites**: Preprocessed structure libraries (including CF files) are available at our website, but we describe the detailed procedures used to build the libraries for user information.
   (a) *Download PDB files*: Users can download all of small ligand-containing PDB files from the Protein Data Bank (PDB, http://www.rcsb.org) [20]. To obtain a list of PDB files to be downloaded, users can use the "Advanced Search Interface (http://www.rcsb.org/pdb/search/advSearch.do?search=new)" provided by the PDB, which enables queries by specific categories. To build structure libraries, we downloaded the PDB files of X-ray crystallographic structures containing at least one protein and one ligand with resolution of ≤3 Å (i.e., 785,78 PDB files as of February 2016).
   (b) *Build structure library*: Users can use *BuildLigandBS-StructureLibrary* in G-LoSA Toolkit to generate small molecule ligand/BS structure library using the downloaded PDB files. This program assigns BS residues based on ligand molecules identified in the heteroatom section of the PDB file. Heteroatoms having an identical chain ID and sequence number are grouped into one heteroatom group. If a distance of any atom pair from different heteroatom groups is 1–2 Å, the two heteroatom groups are merged into one group. Metal ions, water molecules, small molecular weight ligands (the number of heavy atoms <5), and covalently linked heteroatom groups to a protein are removed from the heteroatom group set. Each remaining heteroatom group is assigned to a ligand. If any atom of a residue in a protein is within 4.5 Å of its cognate ligand, the residue is defined as the BS

residue. Metal ion- and water-mediated interactions with a ligand are also considered to define the BS residues using cutoff distances of 3.2 Å and 3.0 Å, respectively. A set of the BS residues is referred to as a BS structure. The *BuildLigandBSStructureLibrary* program generates output structure files in PDB format for identified ligands and BS as well as the corresponding proteins for each PDB file in a given list.

(c) *Remove redundancy in structure library*: It is recommended to remove structural redundancy in a generated ligand/BS structure library, aiming at reducing computational time for G-LoSA searches against the library. G-LoSA Toolkit provides the *RemoveRedundantLigandBS* program for this task. This performs filtering of redundant BS for a given list based on sequence identity between protein chains and structure similarity between ligands. The program requires a list with two columns (BS structures and resolution) and a BLAST clustering result file for protein chains in the PDB (http://www.rcsb.org/pdb/statistics/clusterStatistics.do) as inputs. The resolutions for each BS can be added by interrogating its original PDB file using the *AddPropertyResolution* program. Open Babel (http://openbabel.org) [21] needs to be installed for FP2 fingerprint-based ligand structure comparison quantified by the Tanimoto coefficient. When a homologous BS pair is detected, the application removes a BS structure with lower resolution from the list.

(d) *Generate chemical feature files*: CF files for each BS structure in a structure library are needed to run G-LoSA. For a list of BS structures, the user can generate the CF files using the *GenerateCFFiles* program in G-LoSA Toolkit.

3. **PDB format**: This is a standard data format (http://www.wwpdb.org/documentation/file-format) to represent experimentally determined 3D structures of biological macromolecules in the PDB. For its input and output structures, G-LoSA uses the PDB structure file format. CF file is also based on this format. G-LoSA recognizes the end of an input structure by "TER" record. Therefore, each input structure file for G-LoSA must end with this record.

4. **Positional overlap Tanimoto coefficient (T_{PO})**: Positional overlap Tanimoto coefficient between template ligands is calculated by

$$T_{PO} = \frac{N_O}{N_1 + N_2 + N_O} \tag{2}$$

where N_1 and N_2 are the number of non-hydrogen atoms in two template ligands, and N_O is the number of overlapped atoms. If the distance between an atom of the first template ligand and its nearest atom of the second template ligand is ≤1.2 Å, the atom of the first template is defined as an overlapped atom.

Acknowledgments

This work was supported by NIH U54GM087519, KU GRF2301048, and XSEDE MCB070009.

References

1. Schnoes AM, Brown SD, Dodevski I, Babbitt PC (2009) Annotation error in public databases: misannotation of molecular function in enzyme superfamilies. PLoS Comput Biol 5 (12):e1000605
2. Liolios K, Chen IMA, Mavromatis K, Tavernarakis N, Hugenholtz P, Markowitz VM, Kyrpides NC (2010) The Genomes On Line Database (GOLD) in 2009: status of genomic and metagenomic projects and their associated metadata. Nucleic Acids Res 38:D346–D354
3. Rost B (2002) Enzyme function less conserved than anticipated. J Mol Biol 318(2):595–608
4. Chandonia JM, Brenner SE (2006) The impact of structural genomics: expectations and outcomes. Science 311(5759):347–351
5. Skolnick J, Zhou HY, Gao M (2013) Are predicted protein structures of any value for binding site prediction and virtual ligand screening? Curr Opin Struct Biol 23(2):191–197
6. Konc J, Janezic D (2014) Binding site comparison for function prediction and pharmaceutical discovery. Curr Opin Struct Biol 25:34–39
7. Shumilin IA, Cymborowski M, Chertihin O, Jha KN, Herr JC, Lesley SA, Joachimiak A, Minor W (2012) Identification of unknown protein function using metabolite cocktail screening. Structure 20(10):1715–1725
8. Rose PW, Beran B, Bi C, Bluhm WF, Dimitropoulos D, Goodsell DS, Prlic A, Quesada M, Quinn GB, Westbrook JD, Young J, Yukich B, Zardecki C, Berman HM, Bourne PE (2011) The RCSB Protein Data Bank: redesigned web site and web services. Nucleic Acids Res 39 (Database issue):D392–D401
9. Brylinski M, Skolnick J (2008) A threading-based method (FINDSITE) for ligand-binding site prediction and functional annotation. Proc Natl Acad Sci U S A 105(1):129–134
10. Lee HS, Zhang Y (2012) BSP-SLIM: a blind low-resolution ligand-protein docking approach using predicted protein structures. Proteins 80(1):93–110
11. Andreeva A, Howorth D, Brenner SE, Hubbard TJ, Chothia C, Murzin AG (2004) SCOP database in 2004: refinements integrate structure and sequence family data. Nucleic Acids Res 32(Database issue):D226–D229
12. Russell RB (1998) Detection of protein three-dimensional side-chain patterns: new examples of convergent evolution. J Mol Biol 279 (5):1211–1227
13. Lee HS, Im W (2012) Identification of ligand templates using local structure alignment for structure-based drug design. J Chem Inf Model 52(10):2784–2795
14. Lee HS, Im W (2013) Ligand binding site detection by local structure alignment and its performance complementarity. J Chem Inf Model 53(9):2462–2470
15. Lee HS, Im W (2016) G-LoSA: an efficient computational tool for local structure-centric biological studies and drug design. Protein Sci 25(4):865–876
16. Li B, Turuvekere S, Agrawal M, La D, Ramani K, Kihara D (2008) Characterization of local geometry of protein surfaces with the visibility criterion. Proteins 71(2):670–683
17. Greene LH, Lewis TE, Addou S, Cuff A, Dallman T, Dibley M, Redfern O, Pearl F, Nambudiry R, Reid A, Sillitoe I, Yeats C, Thornton JM, Orengo CA (2007) The CATH domain structure database: new protocols and classification levels give a more comprehensive resource for exploring evolution. Nucleic Acids Res 35(Database issue):D291–D297
18. Finn RD, Coggill P, Eberhardt RY, Eddy SR, Mistry J, Mitchell AL, Potter SC, Punta M,

Qureshi M, Sangrador-Vegas A, Salazar GA, Tate J, Bateman A (2016) The Pfam protein families database: towards a more sustainable future. Nucleic Acids Res 44(D1):D279–D285

19. Tauchert MJ, Fourmann JB, Christian H, Luhrmann R, Ficner R (2016) Structural and functional analysis of the RNA helicase Prp43 from the thermophilic eukaryote Chaetomium thermophilum. Acta Crystallogr F Struct Biol Commun 72(Pt 2):112–120

20. Berman HM, Westbrook J, Feng Z, Gilliland G, Bhat TN, Weissig H, Shindyalov IN, Bourne PE (2000) The Protein Data Bank. Nucleic Acids Res 28(1):235–242

21. O'Boyle NM, Banck M, James CA, Morley C, Vandermeersch T, Hutchison GR (2011) Open Babel: an open chemical toolbox. J Cheminform 3:33

Chapter 9

Local Alignment of Ligand Binding Sites in Proteins for Polypharmacology and Drug Repositioning

Michal Brylinski

Abstract

The administration of drugs is a key strategy in pharmacotherapy to treat diseases. Drugs are typically developed to modulate the function of specific proteins, which are directly associated with particular disease states. Nonetheless, recent studies suggest that protein-drug interactions are rather promiscuous and the majority of pharmaceuticals exhibit activity against multiple, often unrelated proteins. Certainly, the lack of selectivity often leads to drug side effects; on the other hand, these polypharmacological attributes can be used to develop drugs acting on multiple targets within a unique disease pathway, as well as to identify new targets for existing drugs, which is known as drug repositioning. To support drug development and repurposing, we developed *e*MatchSite, a new approach to detect those binding sites having the capability to bind similar compounds. *e*MatchSite is available as a standalone software and a webserver at http://www.brylinski.org/ematchsite.

Keywords *e*MatchSite, Drug development, Drug repositioning, Polypharmacology, Computer-aided drug discovery, Binding site alignment, Sequence order-independent alignment

1 Introduction

Network analysis of interactions between proteins and small organic compounds is broadly applicable throughout the drug development process in both biology and chemistry. The classical picture of selective ligand binding has been challenged by experimental and computational studies, which strongly suggest that the space of protein-ligand interactions is dense and highly connected [1]. Several independent studies were conducted to estimate the promiscuity of protein–ligand interactions. For example, a large-scale across-target activity analysis carried out for 189,807 active compounds from PubChem [2] shows that the majority (62%) of them exhibit activity against multiple, often unrelated targets [3]. Another study investigating a set of 3138 compounds against 79 targets reported that 47% of the compounds can be classified as "promiscuous" and 24% as "highly promiscuous" with multiple

Daisuke Kihara (ed.), *Protein Function Prediction: Methods and Protocols*, Methods in Molecular Biology, vol. 1611,
DOI 10.1007/978-1-4939-7015-5_9,

targets hit at an IC_{50} of <10 μM [4]. Furthermore, a thorough survey of a network of 5215 protein-ligand interactions connecting 829 compounds with 557 targets estimated that the average number of target proteins per ligand is 6.3 [5]. Although this notable binding promiscuity may complicate drug development, it also creates appealing opportunities for polypharmacology and drug repurposing.

Classical algorithms detecting relationships between proteins widely used in bioinformatics cannot be applied to investigate drug cross-reactivity because many compounds bind to multiple proteins that are totally unrelated to each other at the global sequence and structure levels [6, 7]. Therefore, a comprehensive analysis of the protein-drug interaction space requires a different set of tools. A direct comparison of binding sites is capable of describing ligand binding at the molecular level to provide useful insights into the compound mode of action [8]. Most algorithms for binding site matching fall into two categories: alignment-free and alignment-based methods. Geometric hashing is a typical example of the alignment-free approach; it measures the overall similarity of two binding sites, however, without providing any structural information on putative ligand binding modes and molecular interactions with target proteins [9]. In contrast, methods based on binding site alignments elucidate why two sites are similar, identify the sets of atoms/residues that contribute to the similarity, and describe putative ligand binding modes. SuMo (Surfing the Molecules) was one of the first approaches to use a residue-independent stereochemical group description combined with a fast, graph-based algorithm to compare protein structures and substructures [10]. Another method, SiteEngine, matches low-resolution protein surfaces constructed by converting triangles of physicochemical properties into a discrete set of chemically important points [11]. Finally, SOIPPA performs sequence order-independent profile-profile alignments of binding pockets using a coarse-grained representation of protein structures [12].

Despite encouraging progress in the development of sequence order-independent algorithms for ligand binding site alignment, many of these approaches perform well only against high-quality binding sites extracted from experimental protein structures. This insufficient accuracy hinders the reconstruction of protein-drug interaction networks across proteomes; thus, it is imperative to develop new approaches insensitive to structural deformation in ligand binding regions of protein models. To mitigate this issue, we developed *e*MatchSite, a new algorithm that performs sequence order-independent local binding site alignments using computer-generated protein models [13]. A key feature responsible for its high performance is the extensive use of evolutionary information that can be extracted even from weakly homologous templates complexed with ligands. In addition, *e*MatchSite provides a calibrated significance score to identify those pockets capable of binding chemically similar ligands regardless of any global sequence and structure similarities between the target proteins. Benchmarking calculations demonstrate that *e*MatchSite outperforms other algorithms constructing sequence order-independent alignments of ligand binding sites. Importantly, *e*MatchSite maintains its high prediction accuracy against protein models; therefore, it opens up the possibility of investigating drug-protein interactions for complete proteomes with prospective systems-level applications in polypharmacology and rational drug repositioning.

2 Materials

2.1 Input Data

Input data for *e*MatchSite consist of two protein structures in the Protein Data Bank (PDB) format, whose ligand binding sites were annotated by *e*FindSite (*see* **Note 1**). *e*FindSite is a ligand binding site prediction and virtual screening algorithm that detects common ligand binding sites in a set of evolutionarily related proteins identified by meta-threading [14, 15]. In order to perform binding site annotation with *e*FindSite, users can employ either its standalone version or webserver located at http://www.brylinski.org/efindsite (*see* **Note 2**). It is noteworthy that both *e*FindSite and *e*MatchSite work well not only with experimentally solved structures, but also with computer-generated protein models (*see* **Note 3**).

2.2 Programs Used

*e*MatchSite is written in C++ and requires the following libraries: zlib (www.zlib.net), gzstream (www.cs.unc.edu/Research/compgeom/gzstream), and libsvm (www.csie.ntu.edu.tw/~cjlin/libsvm). In addition, *e*MatchSite requires a compound library, which is available at http://www.brylinski.org/content/ematchsite-standalone-package. Below, we describe options for running the standalone version of *e*MatchSite.

2.2.1 Input Options

`-i input_file`, where `input_file` is a single text file providing the location of all data files required by *e*MatchSite. Each line should contain only one keyword followed by a space and the path to the input file. Lines starting with # are ignored. List of keywords (`A`—first protein, `B`—second protein):

- `structureA` and `structureB`—target structures in the PDB format
- `profilesA` and `profilesB`—sequence profiles
- `secstrA` and `secstrB`—secondary structure profiles
- `pocketsA` and `pocketsB`—*e*FindSite pockets
- `numberA` and `numberB`—*e*FindSite pocket numbers (default 1)

- `alignmentsA` and `alignmentsB`—*e*FindSite alignments
- `ligandsA` and `ligandsB`—*e*FindSite ligands

Alternatively, users can specify the path to individual data files from command line. For example, the following arguments can be used to provide the location of target structures:

`-structureA first_pdb`, where `first_pdb` is the first protein in the PDB format,

`-structureB second_pdb`, where `second_pdb` is the second protein in the PDB format.

The path to other input data can be specified in a similar way.

2.2.2 Output Options

`-o output_name`, where `output_name` is a PDB file containing all results from *e*MatchSite.

2.2.3 Virtual Screening Options

`-m score_func`, where `score_func` is a scoring function to perform ligand-based virtual screening. Currently, implemented functions include single and combined scores (*see* [15] for details):

Single scoring functions:

`tst`—classical Tanimoto coefficient for Daylight fingerprints,

`tsa`—average Tanimoto coefficient for Daylight fingerprints,

`tsc`—continuous Tanimoto coefficient for Daylight fingerprints,

`tmt`—classical Tanimoto coefficient for MACCS fingerprints,

`tma`—average Tanimoto coefficient for MACCS fingerprints,

`tmc`—continuous Tanimoto coefficient for MACCS fingerprints.

Combined scoring functions:

`sum`—data fusion using the sum rule (default),

`max`—data fusion using the max rule,

`min`—data fusion using the min rule,

`svm`—machine learning using Support Vector Machines.

2.2.4 Output Files

*e*MatchSite outputs a single file that contains (1) numerical scores for the constructed alignment of binding sites, (2) aligned residue pairs with the corresponding Cα-Cα distances, (3) transformation matrices to superpose the second protein onto the first protein, and (4) the coordinates of the second protein upon the superposition of two binding sites.

2.3 Web Sites

The webserver available at http://www.brylinski.org/ematchsite provides a convenient interface to run *e*MatchSite. In addition, the website provides a standalone package that can be installed locally for high-throughput computations, benchmarking datasets and results for an easy comparison with other algorithms constructing local alignments of binding sites, as well as a detailed manual and tutorial to help run *e*MatchSite.

3 Methods

3.1 Web Submission Form

The submission form for the *e*MatchSite webserver requires two target pockets annotated by *e*FindSite. Fig. 1 shows that users need to provide a 10-digit *e*FindSite ticket for each target and specify the pocket number if multiple pockets are identified in the target structure. In addition, each submission requires a unique "Job ID" and, optionally, an email address where the results will be sent. If a user prefers not to use email notifications, the automatically generated 10-digit ticket for each submission can be used to check the job status using the "Job Tracking" form in the right sidebar at http://www.brylinski.org.

3.2 Result Page

Results generated by the *e*MatchSite webserver are arranged into several sections, as shown in Fig. 2. The first section (Fig. 2a) gives the general information including the "Job ID," the "*e*MatchSite ticket" that can be used to retrieve the results within a month from the submission date, and the identifiers of protein targets. Further, three numerical scores for the local alignment of binding sites are provided, including the alignment score ranging from 0 to 1, the

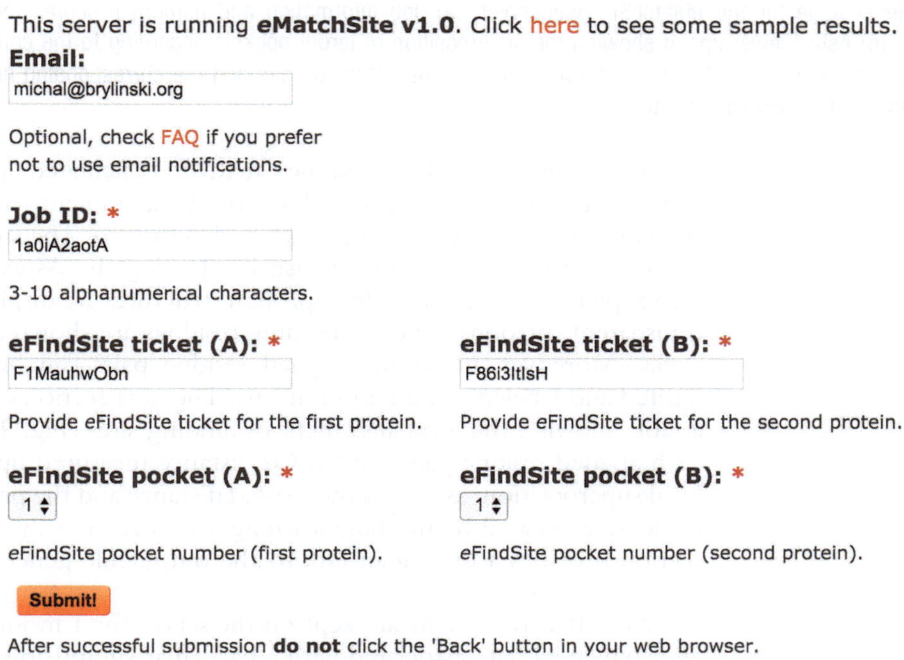
This server is running **eMatchSite v1.0**. Click here to see some sample results.

Email:

michal@brylinski.org

Optional, check FAQ if you prefer not to use email notifications.

Job ID: *

1a0iA2aotA

3-10 alphanumerical characters.

eFindSite ticket (A): *

F1MauhwObn

Provide eFindSite ticket for the first protein.

eFindSite ticket (B): *

F86i3ltlsH

Provide eFindSite ticket for the second protein.

eFindSite pocket (A): *

1

eFindSite pocket number (first protein).

eFindSite pocket (B): *

1

eFindSite pocket number (second protein).

Submit!

After successful submission **do not** click the 'Back' button in your web browser.

Fig. 1 Web submission form for the *e*MatchSite server. Fields marked with *red asterisks* are mandatory

a **eMatchSite Results**

eMatchSite

Job ID:	adenine
eMatchSite ticket:	Mamlasiaty
First protein:	2aotA (pocket 1)
Second protein:	1a0iA (pocket 1)
Alignment score:	0.980
Residues aligned:	12
Cα-RMSD:	4.048Å
These pockets are:	similar

b **Local structure alignment:**

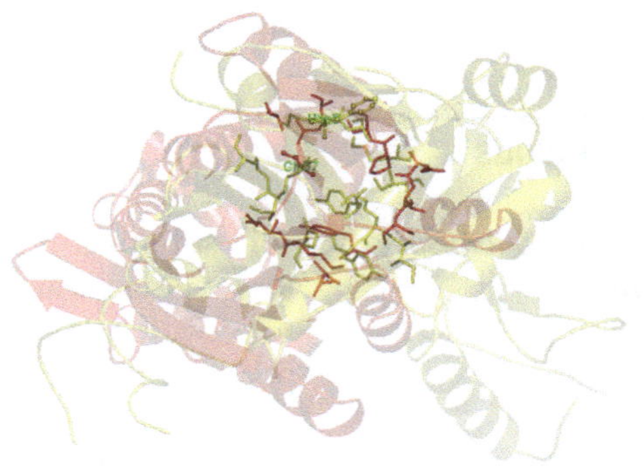

astex

Colors: 2aotA (red), 1a0iA (yellow)

Highlight residue pairs:

8↔57, 9↔90, 10↔87, 32↔115, 33↔143, 34↔141, 38↔138, 54↔59, 92↔137, 140↔116, 211↔114, 229↔86

Controls:

LeftMouse Rotate
Shift+LeftMouse Zoom
Ctrl+LeftMouse Translate

c **Aligned residues:**

2aotA	↔ 1a0iA	Distance [Å][a]	SVR score[b]	SVC score[c]
8 F	57 G	5.431	6.787	0.0695
9 K	90 Q	2.634	3.869	0.0771
10 A	87 S	1.734	5.159	0.0609
32 I	115 T	4.951	7.207	0.0672
33 K	143 Y	3.958	5.896	0.0629
34 Y	141 L	3.154	4.864	0.0639
38 R	138 I	2.290	4.253	0.0599
54 R	59 A	5.608	7.408	0.0615
92 E	137 M	1.771	5.008	0.0870
140 Y	116 S	6.025	4.526	0.0657
211 I	114 E	3.749	7.339	0.0698
229 K	86 P	4.111	8.870	0.0690

[a] Cα-Cα distance measured upon local superposition; [b] Cα-Cα distance in Å estimated by Support Vector Regression; [c] Alignment probability from Support Vector Classifier.

d **Download results:**

eMatchSite output file: 2aotA.eMatchSite.pdb.gz

Fig. 2 Result page for the *e*MatchSite webserver. (**a**) Job information and numerical scores for pocket similarity, (**b**) AstexViewer applet showing the superposition of target pockets according to the constructed sequence order-independent alignment, (**c**) a list of aligned residue pairs and the corresponding numerical scores, and (**d**) the download section

number of aligned residues, and the Cα-RMSD calculated upon the superposition of target pockets. Based on the alignment score, the similarity of the pair of target pockets is determined. The superposition of target binding sites is visualized using the AstexViewer web applet [16] (Fig. 2b). Here, protein structures are displayed as transparent cartoons, whereas binding residues are shown as solid sticks. Moreover, individual aligned residue pairs can be highlighted and labeled using radio buttons. The next section contains a table showing the local alignment of binding sites (Fig. 2c). For each aligned residue pair, the Cα-Cα distance measured upon the local superposition, as well as the Cα-Cα distance and the probability score estimated by machine learning are listed. Finally, the last section provides a download link to the output file generated by *e*MatchSite (Fig. 2d).

Note that the results are kept on the server for 1 month only, after which all data associated with a particular submission will be deleted. However, we keep one sample job for each webserver, so that users can quickly find out whether the webservers offer a desired functionality. These sample results can be accessed anytime either by clicking on links provided in the submission web forms, e. g., "This server is running ***e*MatchSite v1.0**. Click here to see some sample results," or by using 10-digit tickets, "Futermylok" for *e*FindSite and "Mamlasiaty" for *e*MatchSite, in the "Job Tracking" form in the right sidebar.

3.3 Case Studies

We selected a couple of illustrative examples to demonstrate how *e*MatchSite detects those pockets binding similar ligands in non-homologous proteins by constructing the sequence-order independent alignments of their binding sites. The primary target is benzoylformate decarboxylase (BFD) from *Pseudomonas putida* complexed with thiamin-2-thiazolone diphosphate (PDB-ID: 1yno) [17]. BFD belongs to the family of enzymes dependent on thiamine diphosphate and catalyzes the conversion of benzoylformate to benzaldehyde and carbon dioxide. Thiamin-2-thiazolone diphosphate (ThTDP) is a potent inhibitor of several thiamin-dependent enzymes that initiate the catalyzed reactions by forming a covalent adduct between the substrate and thiamin diphosphate (ThDP) through the C2 atom of the thiazolium ring [18]. In ThTDP, the proton on C2 is replaced with an oxygen atom to effectively inactivate thiamin-dependent enzymes. ThTDP binds to its target enzymes with an essentially identical binding mode as ThDP, but at a 10–1000 higher affinity compared to ThDP [19–21].

In addition to the primary target BFD, we selected two off-targets known to bind ThTDP, oxalyl-coenzyme A decarboxylase from *Oxalobacter formigenes* (OXC, PDB-ID: 2c31) [22] and the dehydrogenase/decarboxylase component of the human branched-chain α-ketoacid dehydrogenase complex (hE1b, PDB-ID: 2bff) [23]. OXC plays an important role in the catabolism of the highly toxic compound oxalate and it is structurally similar to BFD with a TM-score [24] of 0.85 despite a low sequence identity of 25% (Table 1, Global similarity). hE1b, which catalyzes the decarboxylation of branched-chain α-ketoacids derived from the amino acids leucine, isoleucine, and valine, shares neither sequence nor structure similarity with BFD (Table 1, Global

Table 1
Global and local structure similarity between the target BFD and off-targets OXC and hE1b

	Global similarity		Local similarity	
Target/off-target	**Sequence identity (%)**	**TM-score**	**Pocket RMSD (Å)**	**Ligand RMSD (Å)**
BFD/OXC	25	0.85	3.18	0.88
BFD/hE1b	25	0.35	2.72	0.89

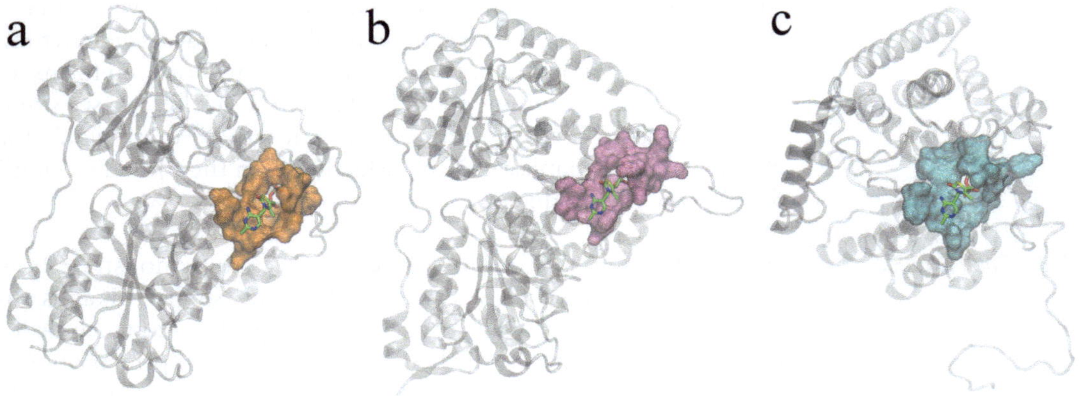

Fig. 3 Ligand binding pockets annotated with *e*FindSite. Target structures are shown as transparent *gray cartoons*, whereas binding residues are rendered as the molecular surface for (**a**) BFD, (**b**) OXC, and (**c**) hE1b. Binding sites in BFD, OXC, and hE1b are colored in *orange*, *pink*, and *cyan*, respectively. ThTDP ligands bound to the target structures are shown as *sticks colored* by the atom type (carbon—*green*, nitrogen—*blue*, oxygen—*red*, sulfur—*yellow*, phosphorus—*tan*)

similarity). Both off-targets represent a challenge to local binding site alignment algorithms due to their low sequence homology with the primary target, BFD.

*e*MatchSite requires binding sites and residues to be annotated by *e*FindSite; therefore, *e*FindSite webserver at http://www.brylinski.org/efindsite was used to identify binding sites in the crystal structures of BFD, OXC, and hE1b. The results are shown in Fig. 3. A binding site for ThTDP in the primary target BFD was identified with a 95% confidence; the Matthews correlation coefficient (MCC) calculated over binding residues is as high as 0.89 (Fig. 3a). The prediction confidence for off-targets OXC and hE1b is 94% and 93%, respectively. MCC calculated for binding residues identified in OXC is 0.78 (Fig. 3b) and 0.88 for hE1b (Fig. 3c). Note that bound ThTDP ligands are shown in Fig. 3 only to assess the accuracy of binding pocket prediction with *e*FindSite, which detects and annotates binding sites in ligand-free protein structures [14, 15].

Despite a low homology between the primary target and off-targets, binding sites in both OXC and hE1b are correctly recognized by *e*MatchSite as highly similar to the ThTDP-binding site in BFD, indicated by a confidence of 93% and 94%, respectively. Using data reported by *e*MatchSite, we can analyze how these high similarity scores were calculated. *e*MatchSite constructs sequence order-independent alignments using machine learning with Support Vector Regression techniques (SVR). Specifically, it assigns SVR scores to all possible combinations of binding residues in the first (target) and the second (off-target) protein. Then, it applies the Kuhn-Munkres algorithm [25, 26] to identify a unique set of residue pairs that give the shortest overall distance between their Cα atoms. The Kuhn-Munkres algorithm, also known as the Hungarian method, belongs to the complexity class P [27], efficiently solving combinatorial assignment problems in polynomial time. In *e*MatchSite, this algorithm produces sequence order-independent alignments whose sum of Cα-Cα distances is guaranteed to be the smallest among all possible alignment combinations.

Clearly, a high correlation between Cα–Cα distances estimated by machine learning (SVR scores) and real distances calculated upon the superposition of bound ThTDP molecules is a critical factor to produce correct alignments. Encouragingly, light gray circles in Fig. 4 demonstrate that the machine learning model implemented in *e*MatchSite accurately predicts Cα-Cα distances; the Pearson correlation coefficients for BFD/OXC (Fig. 4a) and BFD/hE1b (Fig. 4b) are as high as 0.98 and 0.94, respectively. As a consequence, the unique sets of residue pairs selected by the Kuhn-Munkres algorithm to yield the shortest overall Cα-Cα distance actually correspond to the reference alignments constructed by superposing ThTDP molecules bound to the primary target and off-targets (dark gray triangles in Fig. 4). Bear in mind that ThTDP ligands bound to BFD, OXC, and hE1b are used only to validate alignments generated by *e*MatchSite that employs binding pockets annotated by *e*FindSite in ligand-free target structures.

Geometrical and physicochemical matching of binding sites in *e*MatchSite is supported by a chemical correlation, which was originally devised to study the inhibitor cross-reactivity within the human kinome [28]. In essence, a fingerprint-based virtual screening is performed against two pockets using a nonredundant subset of the ZINC library [29] comprising 23,659 molecules.

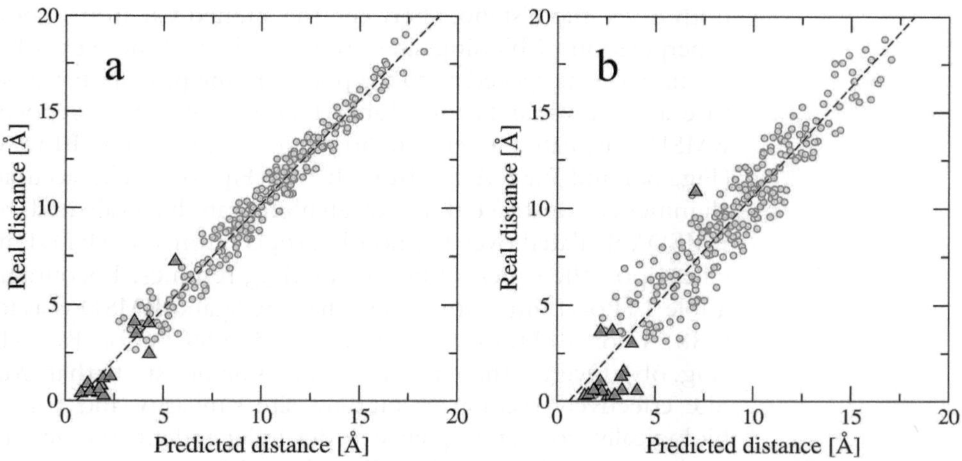

Fig. 4 Accuracy of the prediction of inter-residue distances by *e*MatchSite. Correlation between real distances upon the superposition of ThTDP ligands and those predicted by *e*MatchSite for target protein structures is shown for (**a**) BFD/OXC and (**b**) BFD/hE1b pairs. *Dark gray triangles* show residue pairs from the reference alignment

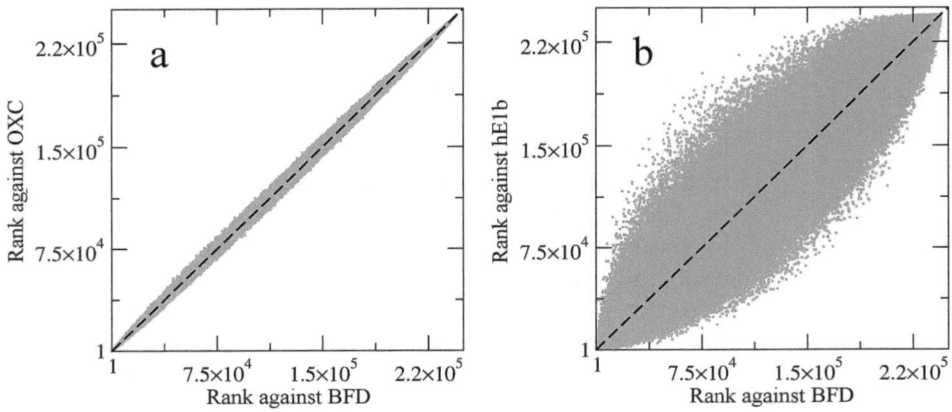

Fig. 5 Chemical correlation between ThTDP binding sites by *e*MatchSite. Rank correlation is plotted for a nonredundant subset of the ZINC library ranked using fingerprint-based virtual screening against a pair of target binding sites, (**a**) BFD/OXC and (**b**) BFD/hE1b

Subsequently, the Kendall τ rank correlation coefficient [30] is calculated for the ranked compounds under the assumption that virtual screening should yield a similar ranking for those pockets binding similar compounds. Indeed, Fig. 5 shows a high chemical correlation between binding sites in the primary and off-targets selected for this case study. The Kendall τ rank correlation coefficient is 0.98 between BFD and OXC (Fig. 5a), and 0.81 between BFD and hE1b (Fig. 5b).

Sequence order-independent alignments constructed by *e*MatchSite for BFD/OXC and BFD/hE1b are reported in Table 2. Sixteen residues are involved in the alignment between BFD and OXC, and 15 residues in that between BFD and hE1b; in both cases, the distances between the aligned Cα atoms upon the superposition of binding sites are fairly short. Binding pockets in off-targets superposed onto the pocket in the primary target structure are shown in Fig. 6. Table 1 (Local similarity) reports the RMSD calculated over Cα atoms of 3.18 Å for BFD/OXC (Fig. 6a) and 2.72 Å for BFD/hE1b (Fig. 6b). The accuracy of alignments constructed by *e*MatchSite can be evaluated by an RMSD calculated over the non-hydrogen atoms of ThTDP molecules upon the superposition of binding residues. Encouragingly, Table 1 (Local similarity) shows that the ligand RMSD is as low as 0.88 Å for BFD/OXC (Fig. 6c) and 0.89 Å for BFD/hE1b (Fig. 6b). Overall, these two case studies demonstrate that *e*MatchSite effectively recognizes binding site similarity and constructs biologically correct sequence order-independent alignments for pockets inferred by *e*FindSite from ligand-free protein structures.

Table 2
Sequence order-independent alignments constructed by *e*MatchSite for BFD/OXC and BFD/hE1b

BFD	OXC		hE1b	
Position	Position	Distance (Å)	Position	Distance (Å)
S375	G394	3.23	–	–
T376	A395	1.07	Y108	3.67
S377	N396	1.52	R109	3.19
G400	G420	2.03	S157	0.43
G401	V421	1.97	P158	0.79
L402	M422	0.61	L159	0.86
G426	G445	0.94	G187	1.19
D427	D446	1.32	E188	1.07
G428	S447	1.38	G189	0.90
S429	A448	0.95	A190	0.71
Y432	F451	0.48	E193	0.62
T456	G479	3.48	Y219	1.57
Y457	G475	2.47	A220	1.34
G458	K478	2.35	N217	5.47
A459	Y477	1.75	H286	3.08
L460	I476	4.31	I221	2.83

Distances between the aligned Cα atoms of BFD and either OXC or hE1b are measured upon the superposition of binding sites

4 Notes

1. Although the current version of *e*MatchSite requires ligand binding sites to be annotated by *e*FindSite, we are working on other prediction protocols to compare binding pockets detected by purely geometrical methods. Nonetheless, *e*FindSite was demonstrated to outperform many other algorithms in large-scale benchmarking calculations; therefore, the combination of *e*FindSite/*e*MatchSite works best in detecting similar binding sites across large datasets of protein structures. Since *e*FindSite typically detects more than one site for the majority of proteins, users may want to specify the binding site of interest if they plan to run *e*MatchSite for a handful of targets. In large-scale applications, the top-ranked binding sites should be used by default because *e*FindSite ranks the best pocket at rank 1 in about 80% of the cases.

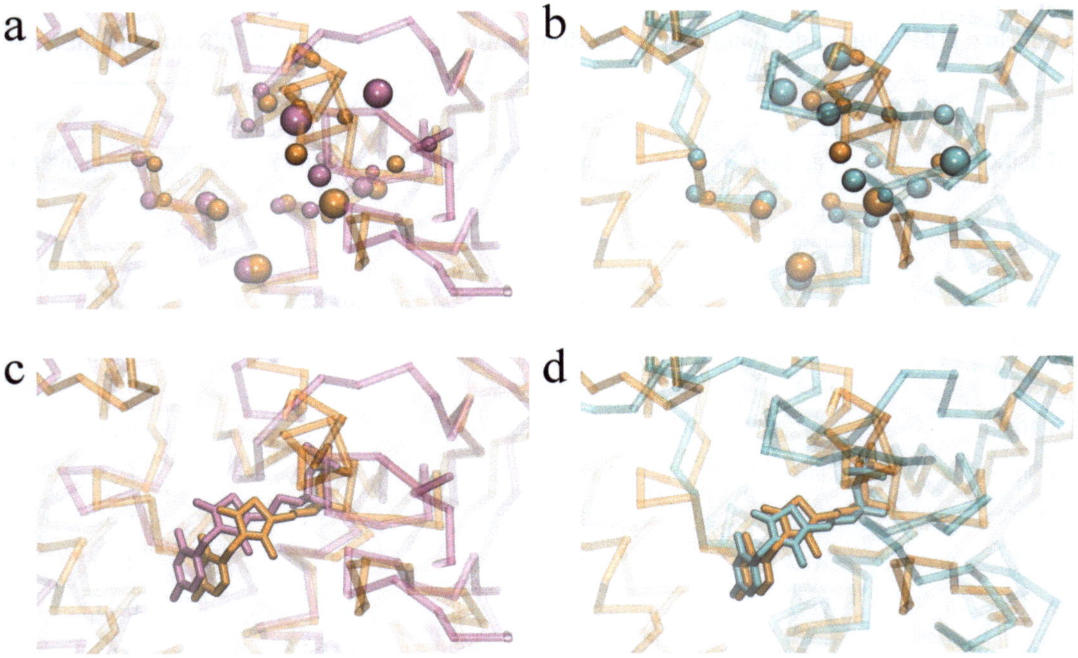

Fig. 6 Sequence order-independent alignments of ThTDP binding sites by *e*MatchSite. (**a**, **b**) Protein structures are superposed according to the local alignment of their binding sites with the Cα atoms of binding residues shown as *solid balls*. (**c**, **d**) Relative orientation of ThTDP ligands upon the local alignment of target binding sites. (**a**, **c**) BFD/OXC and (**b**, **d**) BFD/hE1b. BFD, OXC, and hE1b are colored in *orange*, *pink*, and *cyan*, respectively

2. Web portals for *e*FindSite and *e*MatchSite are intended to study selected proteins and their binding site similarities. However, both tools also have standalone versions that can be installed locally for high-throughput computations across large datasets of protein structures. Our website provides open source codes, the required template libraries, as well as detailed installation instructions and manuals to help users deploy *e*FindSite and *e*MatchSite on their computing systems. It is noteworthy that in addition to a serial code, *e*FindSite was ported to parallel accelerators in order to accelerate binding site annotations using heterogeneous computing systems [31, 32].

3. Both *e*FindSite and *e*MatchSite have been designed to work not only with experimental ligand-bound (holo) and ligand-free (apo) target structures, but also with computer-generated protein models. Compared to crystal structures, the accuracy of *e*FindSite predicting binding residues in high- and moderate-quality structure models decreases only by 4.2% and 9.9%, respectively [14]. Similarly, *e*MatchSite also maintains its capability to construct highly accurate alignments when protein models are used. Depending on the model quality, the

percentage of correctly aligned binding sites is only 4–9% lower than those aligned using crystal structures [13]. On that account, binding site similarities can be effectively detected using homology models generated across proteomes by contemporary protein structure prediction techniques.

Acknowledgments

The research reported in this publication was supported by the National Institute of General Medical Sciences of the National Institutes of Health under Award Number R35GM119524. The content is solely the responsibility of the authors and does not necessarily represent the official views of the National Institutes of Health.

References

1. Paolini GV, Shapland RH, van Hoorn WP, Mason JS, Hopkins AL (2006) Global mapping of pharmacological space. Nat Biotechnol 24:805–815
2. Wang Y, Bolton E, Dracheva S, Karapetyan K, Shoemaker BA, Suzek TO, Wang J, Xiao J, Zhang J, Bryant SH (2010) An overview of the PubChem BioAssay resource. Nucleic Acids Res 38:D255–D266
3. Li Q, Cheng T, Wang Y, Bryant SH (2010) PubChem as a public resource for drug discovery. Drug Discov Today 15:1052–1057
4. Azzaoui K, Hamon J, Faller B, Whitebread S, Jacoby E, Bender A, Jenkins JL, Urban L (2007) Modeling promiscuity based on in vitro safety pharmacology profiling data. ChemMedChem 2:874–880
5. Mestres J, Gregori-Puigjane E, Valverde S, Sole RV (2008) Data completeness—the Achilles heel of drug-target networks. Nat Biotechnol 26:983–984
6. Benson SC, Pershadsingh HA, Ho CI, Chittiboyina A, Desai P, Pravenec M, Qi N, Wang J, Avery MA, Kurtz TW (2004) Identification of telmisartan as a unique angiotensin II receptor antagonist with selective PPARgamma-modulating activity. Hypertension 43:993–1002
7. Weber A, Casini A, Heine A, Kuhn D, Supuran CT, Scozzafava A, Klebe G (2004) Unexpected nanomolar inhibition of carbonic anhydrase by COX-2-selective celecoxib: new pharmacological opportunities due to related binding site recognition. J Med Chem 47:550–557
8. Kellenberger E, Schalon C, Rognan D (2008) How to measure the similarity between protein-ligand binding sites? Curr Comput Aided Drug Des 4:209–220
9. Yeturu K, Chandra N (2008) PocketMatch: a new algorithm to compare binding sites in protein structures. BMC Bioinformatics 9:543
10. Jambon M, Imberty A, Deleage G, Geourjon C (2003) A new bioinformatic approach to detect common 3D sites in protein structures. Proteins 52:137–145
11. Shulman-Peleg A, Nussinov R, Wolfson HJ (2004) Recognition of functional sites in protein structures. J Mol Biol 339:607–633
12. Xie L, Bourne PE (2008) Detecting evolutionary relationships across existing fold space, using sequence order-independent profile-profile alignments. Proc Natl Acad Sci U S A 105:5441–5446
13. Brylinski M (2014) eMatchSite: sequence order-independent structure alignments of ligand binding pockets in protein models. PLoS Comput Biol 10:e1003829
14. Brylinski M, Feinstein WP (2013) eFindSite: improved prediction of ligand binding sites in protein models using meta-threading, machine learning and auxiliary ligands. J Comput Aided Mol Des 27:551–567
15. Feinstein WP, Brylinski M (2014) eFindSite: enhanced fingerprint-based virtual screening against predicted ligand binding sites in protein models. Mol Inform 33:135–150
16. Hartshorn MJ (2002) AstexViewer: a visualisation aid for structure-based drug design. J Comput Aided Mol Des 16:871–881
17. Polovnikova ES, McLeish MJ, Sergienko EA, Burgner JT, Anderson NL, Bera AK, Jordan F,

Kenyon GL, Hasson MS (2003) Structural and kinetic analysis of catalysis by a thiamin diphosphate-dependent enzyme, benzoylformate decarboxylase. Biochemistry 42:1820–1830

18. Arjunan P, Chandrasekhar K, Sax M, Brunskill A, Nemeria N, Jordan F, Furey W (2004) Structural determinants of enzyme binding affinity: the E1 component of pyruvate dehydrogenase from *Escherichia coli* in complex with the inhibitor thiamin thiazolone diphosphate. Biochemistry 43:2405–2411
19. Kluger R, Gish G, Kauffman G (1984) Interaction of thiamin diphosphate and thiamin thiazolone diphosphate with wheat germ pyruvate decarboxylase. J Biol Chem 259:8960–8965
20. O'Brien TA, Gennis RB (1980) Studies of the thiamin pyrophosphate binding site of *Escherichia coli* pyruvate oxidase. Evidence for an essential tryptophan residue. J Biol Chem 255:3302–3307
21. Nemeria N, Yan Y, Zhang Z, Brown AM, Arjunan P, Furey W, Guest JR, Jordan F (2001) Inhibition of the *Escherichia coli* pyruvate dehydrogenase complex E1 subunit and its tyrosine 177 variants by thiamin 2-thiazolone and thiamin 2-thiothiazolone diphosphates. Evidence for reversible tight-binding inhibition. J Biol Chem 276:45969–45978
22. Berthold CL, Moussatche P, Richards NG, Lindqvist Y (2005) Structural basis for activation of the thiamin diphosphate-dependent enzyme oxalyl-CoA decarboxylase by adenosine diphosphate. J Biol Chem 280:41645–41654
23. Machius M, Wynn RM, Chuang JL, Li J, Kluger R, Yu D, Tomchick DR, Brautigam CA, Chuang DT (2006) A versatile conformational switch regulates reactivity in human branched-chain alpha-ketoacid dehydrogenase. Structure 14:287–298
24. Zhang Y, Skolnick J (2004) Scoring function for automated assessment of protein structure template quality. Proteins 57:702–710
25. Kuhn HW (1955) The Hungarian method for the assignment problem. Naval Research Logistics Quarterly 2:83–97
26. Munkres J (1957) Algorithms for the assignment and transportation problems. J Soc Ind Appl Math 5:32–38
27. Sipser M (2004) Time Complexity. In: Sipser M (ed) Introduction to the theory of computation. Thompson Learning, Inc., Cambridge, MA
28. Brylinski M, Skolnick J (2010) Cross-reactivity virtual profiling of the human kinome by X-react(KIN): a chemical systems biology approach. Mol Pharm 7:2324–2333
29. Irwin JJ, Shoichet BK (2005) ZINC—a free database of commercially available compounds for virtual screening. J Chem Inf Model 45:177–182
30. Kendall M (1938) A new measure of rank correlation. Biometrika 30:81–89
31. Feinstein W, Moreno J, Jarrell M, Brylinski M (2015) Accelerating the pace of protein functional annotation with Intel Xeon phi coprocessors. IEEE Trans Nanobioscience 14:429–439
32. Feinstein WP, Brylinski M (2015) Accelerated structural bioinformatics for drug discovery. In: Jeffers J, Reinders J (eds) High performance parallelism pearls volume two. Morgan Kaufmann, Waltham, pp 55–72

Chapter 10

WATsite2.0 with PyMOL Plugin: Hydration Site Prediction and Visualization

Ying Yang, Bingjie Hu, and Markus A. Lill

Abstract

Water molecules in the binding site of a protein significantly influence protein structure and function, for example, by mediating protein–ligand interactions or due to water displacement as driving force for ligand binding. The knowledge about location and thermodynamic contributions of binding site water molecules is crucial for understanding protein function. WATsite is a hydration site analysis program that was developed together with an easy-to-use graphical user interface (GUI) based on PyMOL. WATsite identifies hydration sites from a molecular dynamics (MD) simulation trajectory with four different types of explicit water molecules. Hydration sites can be identified with or without the presence of a bound ligand dependent on the scientific problem. The protein desolvation free energy can be estimated for any ligand by summation of the hydration site free energies of the displaced hydration sites. The location and thermodynamic profile of hydration sites mediating the protein-ligand interactions is important for understanding protein-ligand binding. The WATsite program and GUI are available free of charge from http://people.pharmacy.purdue.edu/~mlill/software/watsite/version2.shtml.

Keywords Protein desolvation free energy, Hydration sites, Molecular dynamics

1 Introduction

Water is a crucial participant in virtually all protein functions including protein folding [1–6] and ligand binding [7–10]. Binding site water contributes significantly to the strength of intermolecular interactions in the aqueous phase by mediating protein–ligand interactions, solvating and desolvating both ligand and protein upon protein-ligand dissociation and association [9, 11–14].

In structure-based drug design, a ligand is often modified to displace ordered water molecules in the binding site. Due to the inherent entropic contributions, releasing an ordered water molecule from the binding site into the bulk solvent is thought to be favorable for protein-ligand binding. However, the desolvation free energy of a water molecule can significantly vary depending on the protein and/or ligand environment. For example, a similar

Daisuke Kihara (ed.), *Protein Function Prediction: Methods and Protocols*, Methods in Molecular Biology, vol. 1611,
DOI 10.1007/978-1-4939-7015-5_10,

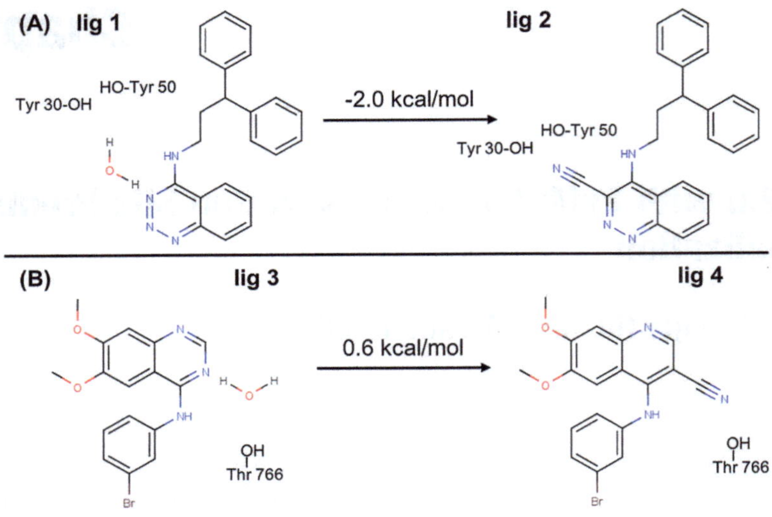

Fig. 1 Example of structure-based drug design by targeting water molecules in the binding site. (**a**) *N*-(3,3-diphenylpropyl)-4-benz-1,2,3-triazine-amine (lig 1) and 3-cyano-*N*-(3,3-diphenylpropyl)-4-cinnolin-amine (lig 2) bound to scytalone dehydratase (SD) in the presence or absence of an interfacial water molecule. Displacement of the interfacial water molecule by the additional cyano group of lig 2 results in a 2.0 kcal/mol gain in binding free energy. (**b**) 4-anilino-6,7-dialkoxyquinazoline inhibitor (lig 3) and 4-anilino-6,7-dialkoxyquinoline-3-carbonitrile inhibitor (lig 4) bound to EGFR kinase. Similar displacement of the interfacial water molecule by the cyano group results in unfavorable binding free energy. This figure is adapted from [8]

strategy was used to displace a single water molecule by a cyano group for two different protein systems: scytalone dehydratase and EGFR kinase (Fig. 1) [10, 15–17].

In Fig. 1a, a water molecule is mediating the interaction between lig 1 and the Tyr-30 and Tyr-50 residues of scytalone dehydratase via hydrogen bonding, while the conversion of the benzotriazine substructure in lig 1 to the 3-cyanocinnoline in lig 2 leads to the displacement of the interfacial water molecule and a 30-fold improvement in K_i value [15]. Similarly, an interfacial water molecule between lig 3 and residue Thr-766 of EGFR kinase is replaced by modifying the quinazoline in lig 3 to 3-cyanoquinoline in lig 4. Despite the same strategy, a threefold decrease in activity was observed for the EGFR kinase system [17]. For a rational decision on whether or not to replace a water molecule by a ligand's functional group, it is therefore important to accurately predict the thermodynamic properties of water molecules in the binding site.

In order to elucidate the thermodynamic profile of individual water molecules and their potential contribution to ligand binding, a hydration site analysis program WATsite was developed together with a graphical user interface based on PyMOL.

2 Materials

2.1 Prerequisites

GROMACS [18], AmberTools [19], reduce [20], and PyMOL [21] have to be installed to utilize WATsite. WATsite has been successfully tested with GROMACS versions 4.5.5, 4.6.1, 5.0.4, and AmberTools 10, 12, 14.

2.2 Installation (on Linux OS)

The WATsite2.0 package can be downloaded at http://people.pharmacy.purdue.edu/~mlill/software/watsite/version2.shtml. *WATsite.tar.gz* needs to be extracted using "tar –zxf WATsite.tar.gz" and the location of WATsite will be noted as $WATSITEHOME. The file WATsite_Settings.txt located in $WATSITEHOME needs to be copied to your home directory.

WATsite plugin can be installed using "Plugin Manager" under Plugin menu in PyMOL. Under the "Install New Plugin" tab, choose *WATsite.py* located under $WATSITEHOME. The HydroEntropy and HydroEnthalpy programs need to be compiled on the platform in the folder $WATSITEHOME using the makefile:

make –f makefile.

2.3 Input Data

The inputs for WATsite are a protein structure and a file defining the binding site. For predicting hydration sites at the protein-ligand interface, a ligand file in mol2 format is required.

2.4 Hydration Site Identification and Desolvation Free Energy Prediction

Here, we briefly describe the individual computational steps of the WATsite procedure and the associated programs. A detailed description of the WATsite method can be found in previous publications [22, 23]. First, the side-chain conformations of ASN, GLN, and HIS, and tautomers and protonation states of HIS residues are adjusted using the program reduce [20]. Next, the protein is solvated in an octahedron of water molecules using a user-specified water model (*see* **Note 1**). A minimum distance of 10 Å between any protein atom and the faces of the octahedron is chosen. Chlorine and sodium ions are then added to neutralize the system. MD simulations are performed using GROMACS [18] with the amber99sb-ildn force field [24]. The system is energy minimized and equilibrated before the production MD simulation. The length of the MD simulations is user-specified (for recommendations, *see* **Note 2**). For the analysis of the production MD trajectory, a 3D grid is placed over the user-defined binding site. The occupancy of the water molecule is distributed onto the 3D grid using all snapshots generated throughout the production run of the MD simulation. The occupancy distribution is then averaged and using a clustering algorithm (*see* **Note 3**) the pronounced peaks are identified to define the hydration site locations. For each hydration site, the programs HydroEntropy and HydroEnthalpy are used to compute the enthalpic (ΔH_{hs}) and entropic (ΔS_{hs}) change of transferring a water molecule from the bulk solvent to the hydration site within the protein binding site.

3 Methods

3.1 Program Setup and Protein System Preparation

The WATsite procedure consists of four steps resembled by the PyMOL plugin as shown in Fig. 2a, three steps for program setup and system preparation, as well as the final hydration site analysis.

1. *Modify settings*: First, the location paths to the required molecular modeling programs (*see* Prerequisites in Subheading 2) need to be correctly specified. Select the menu item "Modify Program Location and Simulation Parameters" from the WATsite menu (Fig. 2a). In the "Modify Settings" tab (Fig. 2b), the correct paths to the programs need to be specified. Users can also specify the number of CPUs provided for running the WATsite simulations.
2. *Modify parameters*: Second, the parameters for the MD simulation and hydration site analysis are specified (Fig. 2c). Based on our previous studies, a 4 ns MD simulation length is typically appropriate for the reliable prediction of the locations and thermodynamic properties of hydration sites [25] (*see* **Note 2**). Currently, four water models (SPC, SPCE, TIP4P, and TIP4P-EW; *see* **Note 1**) and two clustering methods (QT and DBSCAN; *see* **Note 3**) have been tested and can be used for hydration site prediction.
3. *Prepare system*: To specify the protein system, the file containing the protein structure or a structure already displayed in the current PyMOL session is selected (Fig. 2d). To define the protein binding site, a ligand molecule positioned within the binding site needs to be defined, as well as a margin (in Å) for defining the binding pocket enclosing the ligand. The ligand provided in this step is only intended for defining the binding site. It is not included in the MD simulation and the subsequent hydration site identification process. Thus, the user can also construct a "pseudo-ligand" using the binding site residues to define the binding site.

   WATsite allows the identification of hydration sites for both ligand-free (apo) and ligand-bound (holo) protein structures (*see* **Note 4**). If the user intends to predict hydration sites at the interface between the protein and the ligand, the file location of the bound ligand is specified within the PyMOL plugin. The specified ligand will then be included in the MD simulation and the following hydration site identification process. In the current version, no docking service for the user-specified ligand is provided. Therefore, the provided ligand conformation needs to be a meaningful binding pose for the protein.

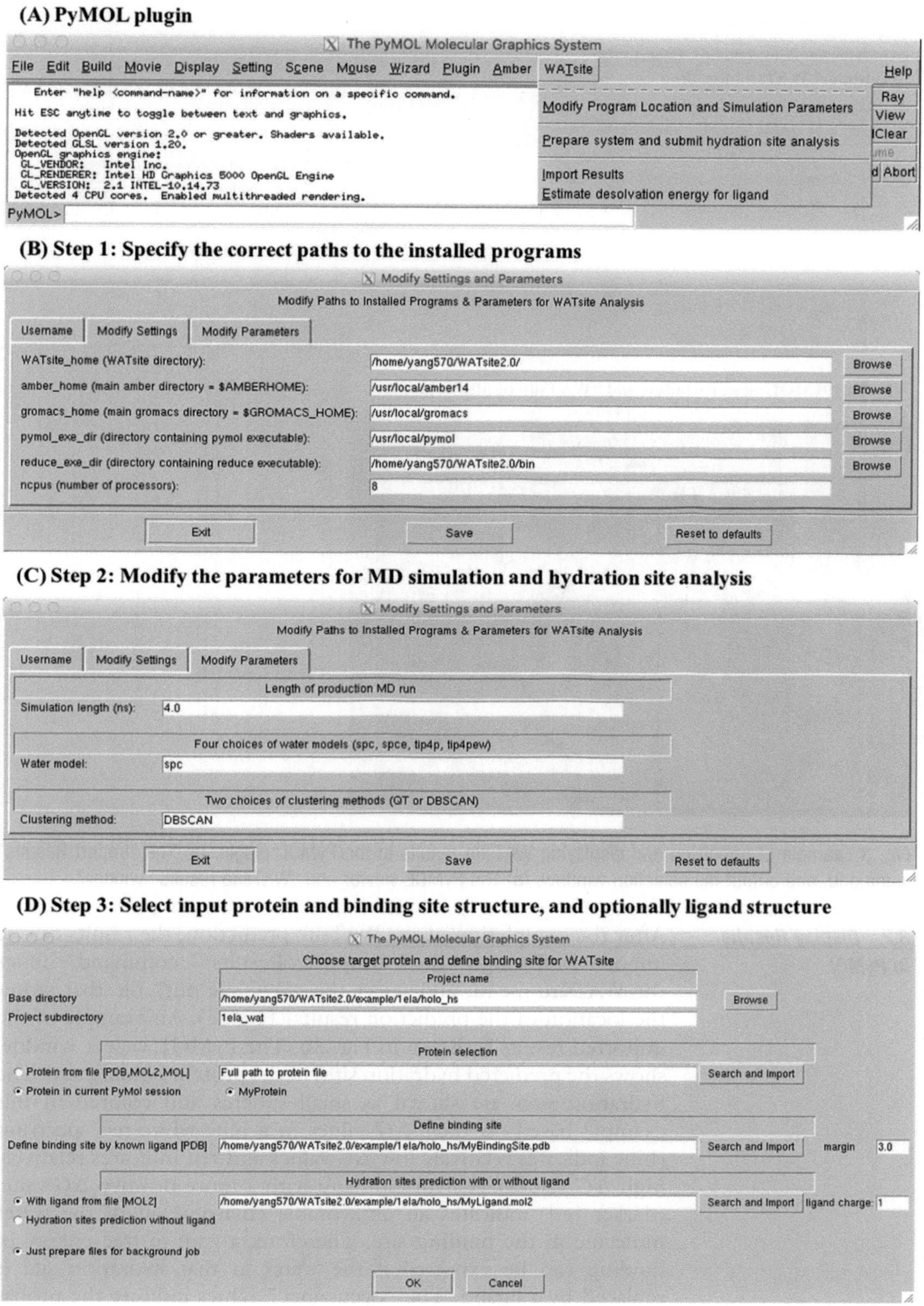

Fig. 2 The steps of hydration site analysis via the PyMOL plugin

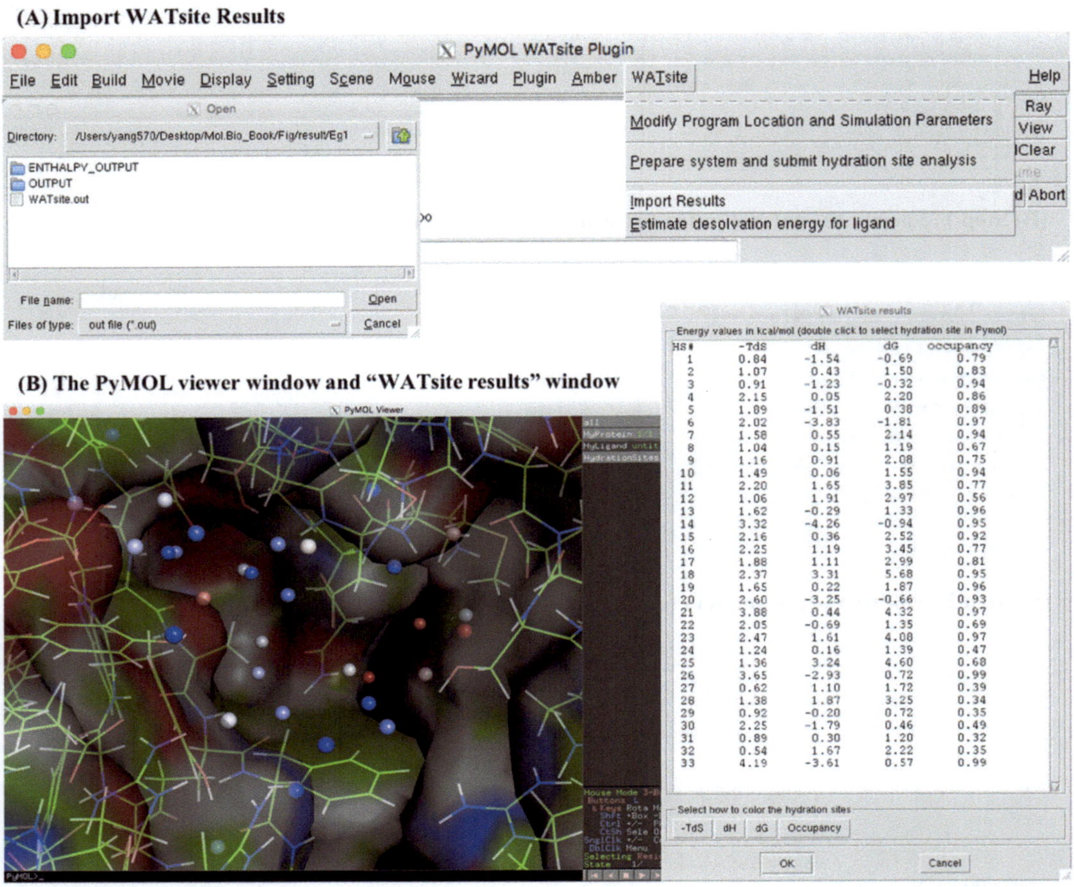

Fig. 3 Example of importing and displaying WATsite results in the PyMOL plugin. (**a**) The "Import Results" command, and output file selection window. (**b**) The PyMOL viewer and "WATsite results" window

3.2 Display Results in PyMOL

After the completion of the WATsite prediction, the results can be imported through the "Import Results" command under the WATsite menu, and select the "WATsite.out" file that stores the location of the prediction results (Fig. 3a). An example of the imported results is shown in Fig. 3b. The PyMOL viewer window shows the predicted hydration sites in the protein binding site. The hydration sites are shown as small spheres and colored in this example based on their ΔG values in a blue-white-red spectrum (blue indicates relatively low ΔG values and red indicates relatively high ΔG values). A hydration site with a more positive ΔG value (darker red) indicates an unfavorable environment of the water molecule in the binding site. Therefore, a gain in free energy of binding can be expected if the water in that hydration site is replaced by a ligand. The "occupancy" values indicate the probability a water molecule is observed in the given hydration site during the MD simulation. The "WATsite results" window lists the estimated desolvation free energy (ΔG), enthalpy (ΔH), entropy ($-T\Delta S$), and occupancy for each hydration site. Hydration sites can be colored according to their "dG," "dH," "$-Td S$," or "Occupancy" values by selecting the corresponding buttons.

3.3 Protein Desolvation Free Energy Estimation

Using the predicted hydration sites and their thermodynamic profile, the protein desolvation free energy can be estimated for bound ligands that replace binding site water molecules. To estimate protein desolvation, the directory containing a set of ligands as well as the radius/cutoff used to select the displaced hydration sites around the ligand atoms need to be specified (Fig. 4a).

For each ligand in the directory, the free energies of hydration sites that are within the user-specified distance to any of the ligand's heavy atoms are summed up to estimate the overall protein desolvation free energy. The predicted desolvation energies (dG, $-TdS$,

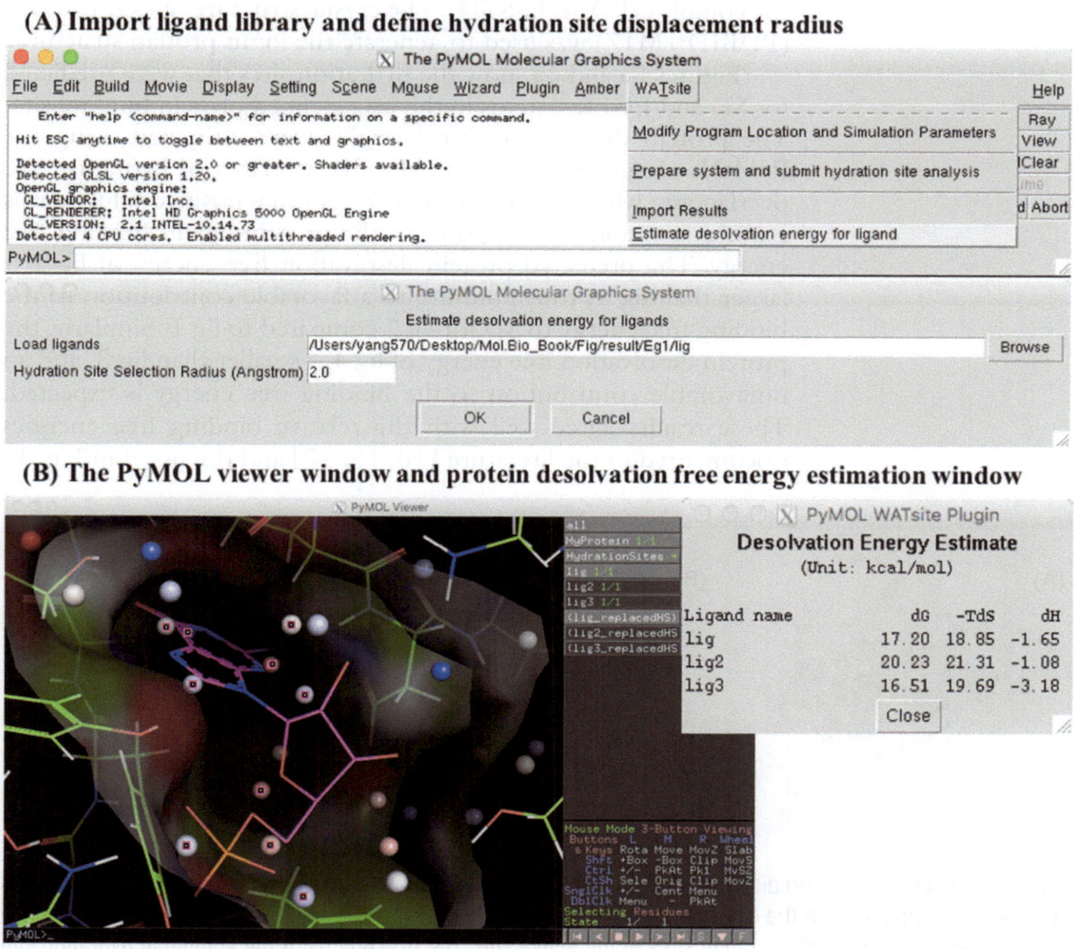

Fig. 4 Example of estimating protein desolvation free energy in the PyMOL plugin. (**a**) The command under WATsite menu, and ligand directory specifying window. (**b**) The PyMOL viewer window and results window

and dH) for each ligand are displayed in a new window, and the selection of displaced hydration sites is highlighted in the PyMOL viewer (Fig. 4b).

4 Case Studies

We show the ability of WATsite to predict the locations of water molecules observed in crystal structures and estimate the protein desolvation free energy of the bound ligands. The hydration sites were predicted for the example systems SD and EGFR mentioned in Subheading 1 (Fig. 1). In order to estimate the difference of water displacement and desolvation free energies between lig 1 & 2 and lig 3 & 4, hydration sites were predicted without the presence of bound ligand. The input pseudo-apo protein structure of SD was generated by removing the bound cyanocinnoline inhibitor from the complex (PDBID:3STD). The crystal structure of EGFR kinase (PDBID:1M17) was used to generate the input protein structure.

Figures 5a and 6a show the hydration sites that were predicted for SD and EGFR systems in the user-defined protein binding site. The protein desolvation free energies were estimated using the PyMOL plugin by adding up the free energies of hydration site overlapping with each ligand (Fig. 7). A more positive value means a more favorable contribution to the protein-ligand binding free energy. The predicted protein desolvation free energy of lig 2 is larger than lig 1, thus pointing to a favorable contribution to the binding free energy to SD for lig 2 compared to lig 1. Similarly, the protein desolvation free energy of lig 4 is smaller than lig 3, and an unfavorable contribution to the binding free energy is expected. These results agree well with the relative binding free energies documented in the literature [10, 15, 17] and discussed in Fig. 1.

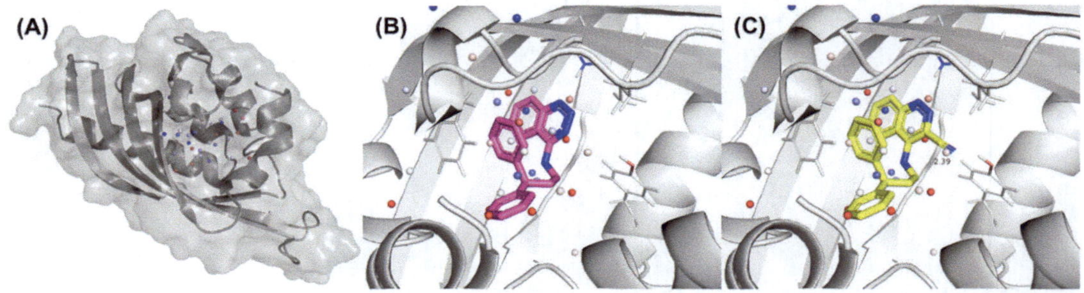

Fig. 5 Representation of predicted hydration sites and bound ligands in scytalone dehydratase (SD). (**a**) Hydration sites predicted in the user-defined binding site. (**b**) Overlay of lig 1 and hydration sites in the active site. (**c**) Overlay of lig 2 and hydration sites in the active site. The free energy of the additional hydration site displaced by lig 2 is labeled

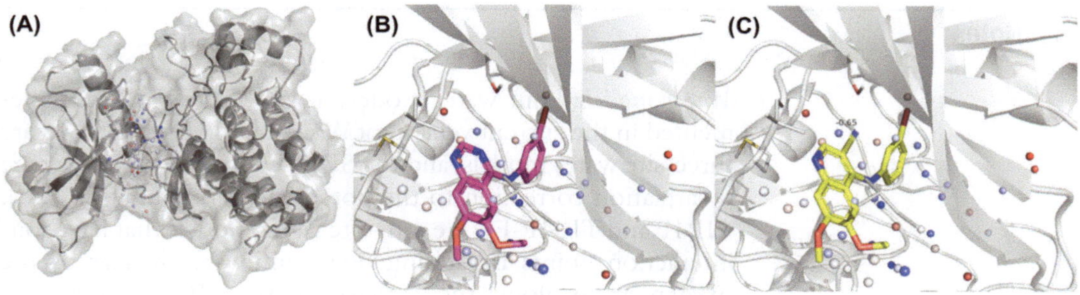

Fig. 6 Representation of predicted hydration sites and bound ligand in EGFR kinase (EGFR). (**a**) Hydration sites predicted in the user-defined binding site. (**b**) Overlay of lig 3 and hydration sites. (**c**) Overlay of lig 4 and hydration sites. The free energy of the additional hydration site displaced by lig 4 is labeled

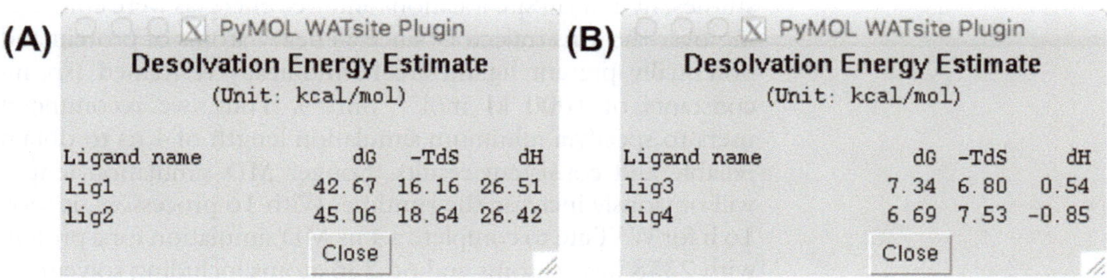

Fig. 7 The protein desolvation free energy estimated for ligands in the cases using the PyMOL plugin. (**a**) The desolvation free energy difference (2.39 kcal/mol) between lig 1 and lig 2 results from one additional hydration site displaced by lig 2. (**b**) The desolvation free energy difference (−0.65 kcal/mol) between lig 3 and lig 4 also results from one additional hydration site displaced by lig 4

Further analysis of the details of hydration sites displacement upon ligand binding reveals that the difference in protein desolvation free energies is actually based on the displacement of one additional water molecule. The modification from lig 1 to lig 2 displaced an additional hydration site with free energy of 2.39 kcal/mol (Fig. 5b, c), while a hydration site with free energy of −0.65 kcal/mol is replaced when lig 3 changes to lig 4 (Fig. 6b, c).

Binding site water displacement is a significant contribution, if not the driving force, of protein-ligand binding. A major application of WATsite is to use the predicted hydration sites to estimate the desolvation free energies involved in replacing binding site water molecules by ligand binding. The above case studies show that accurate estimation of protein desolvation from water displacement is successful in predicting ligands' relative binding affinities. When a lead compound is already known for a specific target, WATsite can be useful in suggesting ligand modification to improve affinity due to the displacement of hydration sites with unfavorable free energies.

5 Notes

1. *Water models*: Four water models have been tested and implemented in the current version of WATsite. SPC and SPC/E are three-site water models, and the SPC/E model adds an average polarization correction to the potential energy function [26]. TIP4P and TIP4P-Ew are four-site water models that have four interaction points by adding one dummy atom near to the oxygen atom along the bisector of the HOH angle. The TIP4P-Ew model was reparameterized for use with Ewald summation methods [27].
2. *Length of MD simulation*: The default production run simulation length is 4 ns based on our previous studies [25]. In those studies, hydration site locations and free energies were converging over the 4 ns trajectory since all heavy atoms of protein and potentially present ligand are harmonically restrained (spring constants of 1000 kJ mol^{-1} nm^{-2}). Thus, we recommend users to specify a minimum simulation length of 4 ns to obtain reliable and consistent results. Longer MD simulation length will obviously increase the runtime. With 16 processors, it took 16 h for WATsite to complete a 4 ns MD simulation for a protein with 2358 heavy atoms and 66,236 atoms including solvent.
3. *Clustering method*: Quality Threshold (QT) and Density-based spatial clustering of applications with noise (DBSCAN) clustering algorithms [28, 29] have been implemented in WATsite. QT clustering method is superior in identifying hydration sites with a large occupancy value during the simulation, whereas DBSCAN clustering method is advantageous in identifying more hydration sites with lower occupancy throughout the simulation.
4. *Hydration site prediction with and without the presence of bound ligand*: The ligand-free (apo) protein structure should be used to predict hydration sites for the computation of the overall protein desolvation free energy upon ligand binding. WATsite also allows for hydration sites prediction at the protein-ligand interface. It should be noted that the partial charge of the ligand is computed using the AM1-BCC method in the current protocol. The method for computing the partial charges of a ligand may significantly influence the desolvation free energy estimation.
5. *Protein desolvation free energy*: It is worth noting that the estimated desolvation free energy only represents the energy of releasing the water molecules from the protein binding site into the bulk solvent. It cannot be used for a direct comparison of the ligands' free energies of binding. The ligand's free energy of binding includes other important contributions such as the

direct protein–ligand interaction energy or desolvation energy of the ligand. The predicted protein desolvation free energy, however, can guide further optimization of a lead compound rationally stabilizing or replacing additional water molecules in the binding site.

Acknowledgments

The authors gratefully acknowledge a grant from the NIH (GM092855) for partially supporting this research.

References

1. Cheung MS, Garcia AE, Onuchic JN (2002) Protein folding mediated by solvation: water expulsion and formation of the hydrophobic core occur after the structural collapse. Proc Natl Acad Sci U S A 99(2):685–690
2. Gao M et al (2010) Water dynamics clue to key residues in protein folding. Biochem Biophys Res Commun 392(1):95–99
3. Kovacs IA, Szalay MS, Csermely P (2005) Water and molecular chaperones act as weak links of protein folding networks: energy landscape and punctuated equilibrium changes point towards a game theory of proteins. FEBS Lett 579(11):2254–2260
4. Sessions RB, Thomas GL, Parker MJ (2004) Water as a conformational editor in protein folding. J Mol Biol 343(4):1125–1133
5. Vajda T, Perczel A (2014) Role of water in protein folding, oligomerization, amyloidosis and miniprotein. J Pept Sci 20(10):747–759
6. Zuo G, Hu J, Fang H (2009) Effect of the ordered water on protein folding: an off-lattice Go-like model study. Phys Rev E Stat Nonlinear Soft Matter Phys 79(3 Pt 1):031925
7. Biela A et al (2012) Water makes the difference: rearrangement of water solvation layer triggers non-additivity of functional group contributions in protein-ligand binding. ChemMedChem 7(8):1423–1434
8. Breiten B et al (2013) Water networks contribute to enthalpy/entropy compensation in protein-ligand binding. J Am Chem Soc 135 (41):15579–15584
9. Li Z, Lazaridis T (2006) Thermodynamics of buried water clusters at a protein-ligand binding interface. J Phys Chem B 110 (3):1464–1475
10. Michel J, Tirado-Rives J, Jorgensen WL (2009) Energetics of displacing water molecules from protein binding sites: consequences for ligand optimization. J Am Chem Soc 131 (42):15403–15411
11. Baron R, Setny P, McCammon JA (2010) Water in cavity-ligand recognition. J Am Chem Soc 132(34):12091–12097
12. Bortolato A et al (2013) Water network perturbation in ligand binding: adenosine A(2A) antagonists as a case study. J Chem Inf Model 53(7):1700–1713
13. Hummer G (2010) Molecular binding: under Water's influence. Nat Chem 2(11):906–907
14. Ladbury JE (1996) Just add water! The effect of water on the specificity of protein-ligand binding sites and its potential application to drug design. Chem Biol 3(12):973–980
15. Chen JM et al (1998) Structure-based design of potent inhibitors of scytalone dehydratase: displacement of a water molecule from the active site. Biochemistry 37(51):17735–17744
16. Gerogiokas G et al (2015) Evaluation of water displacement energetics in protein binding sites with grid cell theory. Phys Chem Chem Phys 17(13):8416–8426
17. Wissner A et al (2000) 4-Anilino-6,7-dialkoxyquinoline-3-carbonitrile inhibitors of epidermal growth factor receptor kinase and their bioisosteric relationship to the 4-anilino-6,7-dialkoxyquinazoline inhibitors. J Med Chem 43(17):3244–3256
18. Pronk S et al (2013) GROMACS 4.5: a high-throughput and highly parallel open source molecular simulation toolkit. Bioinformatics 29(7):845–854
19. Case DA et al (2014) AMBER 14. University of California, San Francisco
20. Word JM et al (1999) Asparagine and glutamine: using hydrogen atom contacts in the choice of side-chain amide orientation. J Mol Biol 285(4):1735–1747

21. The PyMOL Molecular Graphics System, Version 1.8 Schrödinger, LLC
22. Hu B, Lill MA (2012) Protein pharmacophore selection using hydration-site analysis. J Chem Inf Model 52(4):1046–1060
23. Hu B, Lill MA (2014) Watsite: hydration site prediction program with pymol interface. J Comput Chem 35(16):1255–1260
24. Lindorff-Larsen K et al (2010) Improved side-chain torsion potentials for the Amber ff99SB protein force field. Proteins 78(8):1950–1958
25. Yang Y, Hu B, Lill MA (2014) Analysis of factors influencing hydration site prediction based on molecular dynamics simulations. J Chem Inf Model 54(10):2987–2995
26. Zielkiewicz J (2005) Structural properties of water: comparison of the SPC, SPCE, TIP4P, and TIP5P models of water. J Chem Phys 123 (10):104501
27. Horn HW et al (2004) Development of an improved four-site water model for biomolecular simulations: TIP4P-Ew. J Chem Phys 120 (20):9665–9678
28. Ester M et al (1996) A density-based algorithm for discovering clusters in large spatial databases with noise. In: Proceedings of the second international conference on knowledge discovery and data mining (KDD-96). AAAI Press, Menlo Park
29. Heyer LJ, Kruglyak S, Yooseph S (1999) Exploring expression data: identification and analysis of coexpressed genes. Genome Res 9 (11):1106–1115

Chapter 11

Enzyme Annotation and Metabolic Reconstruction Using KEGG

Minoru Kanehisa

Abstract

KEGG is an integrated database resource for linking sequences to biological functions from molecular to higher levels. Knowledge on molecular functions is stored in the KO (KEGG Orthology) database, while cellular- and organism-level functions are represented in the PATHWAY and MODULE databases. Genes in the complete genomes, which are stored in the GENES database, are given KO identifiers by the internal annotation procedure, enabling reconstruction of KEGG pathways and modules for interpretation of higher-level functions. This is possible because all the KEGG pathways and modules are represented as networks of KO nodes. Here we present knowledge-based prediction methods for functional characterization of amino acid sequences using the KEGG resource. Specifically we show how the tools available at the KEGG website including BlastKOALA and KEGG Mapper can be utilized for enzyme annotation and metabolic reconstruction.

Keywords KEGG pathway map, KEGG module, KEGG Mapper, Genome annotation, Pathway analysis, EC number

1 Introduction

With the availability of genome sequences for single organisms and metagenome sequences for multiple organisms in environmental samples, protein function prediction methods can be applied not only to molecular functions of individual proteins but also to cellular-, organism-, and ecosystem-level functions resulting from collective sets of proteins. Among such higher-level functions, metabolism is the most basic biological process. In this article, we focus on metabolism and describe how the KEGG (Kyoto Encyclopedia of Genes and Genomes) database resource [1] can be utilized to predict both individual enzyme functions and overall metabolic functions of an organism or an ecosystem from genome or metagenome sequences.

For more than 50 years, the EC (Enzyme Commission) number system has been developed by the Nomenclature Committee of

Daisuke Kihara (ed.), *Protein Function Prediction: Methods and Protocols*, Methods in Molecular Biology, vol. 1611, DOI 10.1007/978-1-4939-7015-5_11, © Springer Science+Business Media LLC 2017

the International Union of Biochemistry and Molecular Biology (NC-IUBMB) and its predecessors [2]. It is a hierarchical classification of enzymes by the reactions that they catalyze. When an enzymatic reaction is experimentally characterized and reported in a publication, it is given four numbers separated by periods. The first three numbers represent class, subclass, and sub-subclass of a reaction and the fourth is a serial number for an instance under the sub-subclass. The fourth number distinguishes substrate specificity, cofactors, and other variations for the same reaction.

The EC number system is very well established and widely used in many databases, not only to characterize reactions, which is the original intention of this system, but also to characterize enzymes. It is common practice to annotate genes and proteins with EC numbers based on sequence similarity, meaning that EC numbers are treated like ortholog identifiers. When we initiated the KEGG database in 1995, the EC number was used as a primary identifier for reconstruction of metabolic pathways from complete genomes. This was accomplished by drawing metabolic pathway maps with EC number nodes and by annotating enzyme genes in the genome with EC numbers [3]. By 1999, however, problems of EC numbers became apparent, and we started using ortholog IDs, which later became KEGG Orthology (KO) identifiers. Currently, all the KEGG pathway maps, both metabolic and nonmetabolic, are drawn with KO nodes, and genes in the genome are annotated with KOs, enabling reconstruction of pathways and interpretation of higher-level functions. Selected EC numbers are associated with KOs as part of the KO definition.

In general, KOs represent functional orthologs in the context of KEGG pathway maps, but they are also defined as sequence similarity groups. Despite the fact that EC numbers are widely used in sequence databases, the Enzyme Nomenclature list created by NC-IUBMB does not contain sequence information of enzymes. In the KEGG implementation of Enzyme Nomenclature, protein sequences used in the original experiments of characterizing enzymatic functions are manually identified, whenever possible, based on the references given in the ExplorEnz database [2]. These sequences have been used to verify existing KOs and to define new KOs. The knowledge-based prediction methods for metabolic functions presented here are based on this improved KO database and the already well-established KEGG PATHWAY database.

2 Materials

2.1 KO Database

The KEGG GENES database contains gene/protein entries taken mostly from the RefSeq and GenBank databases. The annotation of these entries is done somewhat differently from other sequence databases [1, 4]. They are simply given KO identifiers, also called

K numbers, without rewriting definition or other text information of the original data. The KO database is a collection of functional orthologs, each containing a list of KO-assigned orthologous genes in the GENES database. More importantly, the KO database is a molecular function database. Each KO entry is associated with references and sequence data of experimentally characterized proteins, so that the validity of sequence similarity-based extension can be examined for each ortholog group. The content of the KO database is limited by the content of the GENES database, which used to contain only the complete genomes taken from RefSeq or GenBank. However, it is now possible to include any functionally characterized proteins in the new addendum category of the GENES database.

All the original enzyme sequences that have been identified for the EC number list (*see* Subheading 1) are now fully represented in the GENES database, either in the complete genome category or the addendum category. Existing KOs have been verified and new KOs have been defined accordingly. Note, however, that EC numbers are a collection of instances, i.e., specific enzymes from specific organisms, and the relationships between KOs and ECs are many-to-many. One KO may correspond to multiple ECs and multiple KOs may correspond to a single EC. Furthermore, some EC numbers are not linked to KOs when they are considered unnecessary because of the other existing EC numbers or unfounded because of the lack of supporting sequence data. With the efforts to identify original enzyme sequences, the KO database has become the best curated database for enzyme orthologs.

2.2 Nonredundant GENES Database

As the result of the internal annotation procedure, the KEGG GENES database can be viewed as partitioned into K number groups. This is an advantage over other sequence databases without such partitioning. The sequence similarity search against GENES involves finding most appropriate K number groups, which are directly linked to functional information in the KO database. However, there is a problem that is common to other sequence databases. The GENES database contains a number of redundant sequence data largely due to sequencing of closely related species. For the purpose of KO assignment in the BlastKOALA search described below, nonredundant GENES datasets have been developed by selecting representative sequences of functionally characterized (KO assigned) proteins for each taxonomic group at the species, genus, or family level [1, 4].

In practice this is done as follows. According to the ordering of the KEGG organisms in the NCBI taxonomy (http://www.kegg.jp/brite/br08610), the first genome in each taxonomic group is taken as a representative genome and the remaining genomes are removed except those genes with different K numbers that are not present in the representative genome. This procedure does not fully

Table 1
Nonredundant GENES database files

GENES category	Original sequence data	Nonredundant sequence data generated by	
		Taxonomic grouping only	With CD-HIT clusters
Eukaryotes	eukaryotes.pep	family_eukaryotes.pep genus_eukaryotes.pep	c_family_eukaryotes.pep
Prokaryotes	prokaryotes.pep	genus_prokaryotes.pep species_prokaryotes.pep	c_ genus_prokaryotes.pep
Viruses	T40000.pep		c_viruses.pep
Addendum	T10000.pep	(Included in Eukaryotes or Prokaryotes category)	

(Note) File extension .pep represents a FASTA format amino acid sequence file in KEGG

incorporate diversity of uncharacterized (no KO assigned) proteins. Thus, the dataset for each taxonomic group is supplemented by different CD-HIT clusters [5] for use in the GhostKOALA search, which assigns taxonomic compositions as well as KOs. Table 1 is a summary of currently used nonredundant GENES database files.

2.3 KEGG PATHWAY, MODULE and BRITE Databases

These are the databases for prediction of higher-level functions, in this case, metabolic functions [6]. As of this writing (April 2016), the metabolism section of the KEGG PATHWAY database contains four global maps, 159 regular pathway maps, and five overview maps. Global maps are useful for capturing overall features of reconstructed pathways and for comparison of different organisms or different environmental samples. Regular maps represent higher resolution maps showing more details of individual pathways. While global maps and regular maps provide the same view with different resolutions, overview maps provide a somewhat different view focusing on architectures of metabolic pathways both as chemical and genomic networks [7].

KEGG modules are functional units defined by the logical expression of K numbers. A node in the KEGG metabolic pathway map may be linked to multiple KOs when, for example, an enzyme consists of multiple subunits or taxonomically different groups correspond to the same enzyme. The logical expression used in KEGG modules presents a more precise definition for each KO node and its connection pattern. There are currently 272 KEGG modules corresponding to conserved subpathways in the metabolic pathway, which are a higher resolution representation than regular pathway maps and more directly related to metabolic functions.

The KEGG BRITE database is a collection of hierarchical classifications (ontologies) for genes and proteins, chemical compounds and reactions, drugs and diseases, and organisms. The protein family section of the BRITE database currently contains 11 enzyme families.

2.4 BlastKOALA and KEGG Mapper

KOALA (KEGG Orthology And Links Annotation) is KEGG's internal annotation tool for KO assignment of the GENES database using the result of SSEARCH sequence comparisons [8]. BlastKOALA (www.kegg.jp/blastkoala/) is a web server implementation of the same algorithm for KO assignment after the BLAST search [9] against the nonredundant GENES database. For a given set of amino acid sequences it returns a list of K numbers assigned together with links to KEGG Mapper analysis. There is another KO assignment server, GhostKOALA (www.kegg.jp/ghostkoala/), available at the KEGG website. It uses more rapid GHOSTX [10] rather than BLAST together with a simplified KO assignment scheme. It is suited for assigning KOs to a large amount of metagenomic sequence data and also for finding taxonomic compositions. A brief comparison of the two servers is shown in Table 2.

KEGG Mapper (www.kegg.jp/kegg/mapper.html) is a web service consisting of a collection of KEGG mapping tools. The BlastKOALA or GhostKOALA result is linked to three KEGG mapping tools called Reconstruct Pathway, Reconstruct Module and Reconstruct Brite. These tools reconstruct KEGG pathways, modules, and Brite hierarchies from a given set of K numbers

Table 2
KO assignment programs

	BlastKOALA	GhostKOALA	Annotate sequence
Purpose	Accurate KO assignment of high-quality sequence data, such as complete genomes	Rough KO assignment of a large amount of sequence data, such as metagenomes	Quick KO assignment when related species are present in KEGG
Associated analysis	Reconstruct Pathway Reconstruct Brite Reconstruct Module	Reconstruct Pathway Reconstruct Brite Reconstruct Module Map Taxonomy	Same as BlastKOALA
URL	www.kegg.jp/blastkoala/	www.kegg.jp/ghostkoala/	www.kegg.jp/kegg/tool/annotate_sequence.html
Search program	BLASTP	GHOSTX	BLASTP
Scoring	Weighted sum of BLAST bit scores with a modified KOALA algorithm	Unweighted sum of GHOSTX scores	Same as BlastKOALA
Database	family_eukaryotes.pep genus_eukaryotes.pep genus_prokaryotes.pep species_prokaryotes.pep	c_family_eukaryotes.pep c_ genus_prokaryotes.pep c_viruses.pep	Single family (eukaryotes) Single genus (prokaryotes)
Job	Email-based	Email-based	Interactive

revealing, for example, which types of metabolism are likely to be present in an organism or an environment. In addition, GhostKOALA is linked to another tool called Map Taxonomy, which would assist characterizing taxonomic composition of an environmental sample.

3 Methods

3.1 BlastKOALA Job Submission

Access the BlastKOALA server (www.kegg.jp/blastkoala/) and prepare for the job submission as follows.

1. Enter a query set of amino acid sequences in FASTA format. The maximum size of the query dataset is dependent on the database file size to be searched.
2. Specify taxonomic information of query sequences, if known, which only affects the scoring of KO assignment (*see* **Note 1**).
3. Select a nonredundant GENES database file to be searched (Table 1). There are options to search eukaryotes and prokaryotes separately and to select from different taxonomic ranks. A lower taxonomic rank file is a larger file allowing more accurate KO assignment but taking more time to search.
4. Enter the email address for job submission and accessing result data (*see* **Note 2**).
5. Click on the "Request for email confirmation" button to upload query sequences and other data.

The user will immediately receive an email from the BlastKOALA server. Click on the link in the message to submit the job.

3.2 Accessing Result Data

When the job is completed another notification email is sent. Click on the link in the message to access the result page.

1. The result page contains links for viewing and downloading a list of KO assignments, as well as for performing KEGG Mapper analysis. Figure 1 is an example of the result page, where the pie chart is drawn according to the number of genes assigned to each functional category of the KEGG Orthology system (*see* **Note 3**).
2. From the "Download" link a local file can be created containing the list of query gene names and K numbers assigned. It is recommended to locally store the KO assignment file because all the data for each job are deleted 1 week after the notification email is sent.
3. From the "View" link it is possible to examine more details of the KO assignment. This link may be useful when the downloaded KO assignment file is to be manually improved.

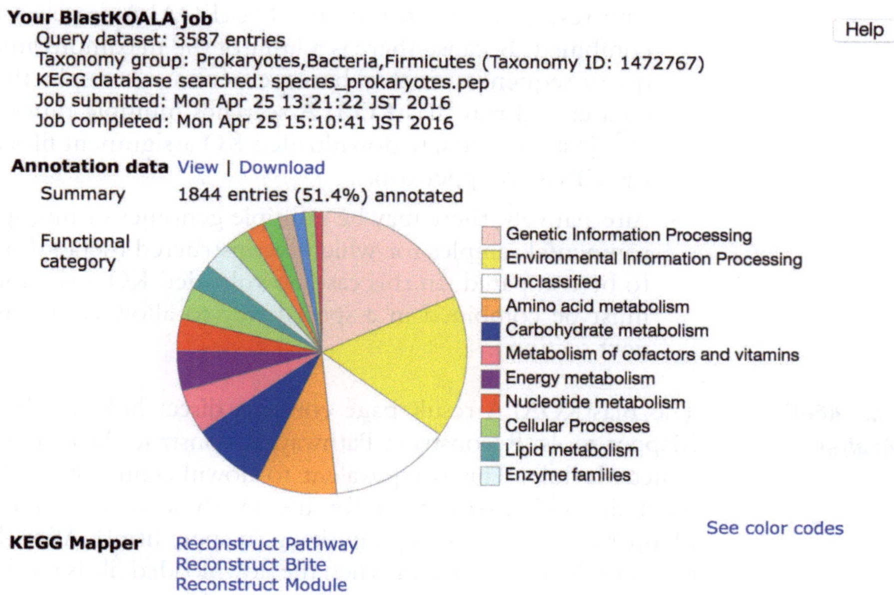

Fig. 1 An example of BlastKOALA result page

4. For the list of enzymes identified, click on the "Reconstruct Brite" link and select "ko01000 Enzymes" under the category of "Protein families: metabolism." The list may be used to convert K numbers to EC numbers if necessary (*see* **Note 4**).
5. For the result of metabolic reconstruction, click on the "Reconstruct Pathway" link and select pathways of interest under "Metabolism."
6. Click also on the "Reconstruct Module" link to examine complete and almost complete modules.

As is apparent from this example, the BlastKOALA service is not limited to enzymes and metabolic pathways. It is up to the user to select proteins and pathways of interest from the result data.

3.3 Manipulating Result Data

The downloaded KO assignment file may be modified or manipulated for further KEGG Mapper analysis.

1. The KO assignment may be improved by manually checking the automatically assigned result. From the "View" link it is possible to examine the second best assignments, which may be used to modify the downloaded file, especially when no KOs are assigned as a first choice. More details of how the first and second choices are determined can be evaluated in a simplified version of the GFIT (Gene Function Identification Tool) table [6], which is linked from each gene name. It contains the BLAST bit score, percent identity, and alignment overlap length, as well as the K number and its definition.

2. The result files from multiple BlastKOALA searches may be combined. Because there is a limit of the maximum amount of query sequences, it may become necessary to split the query dataset and run BlastKOALA searches multiple times. In this case just concatenate downloaded KO assignment files and use the KEGG Mapper tools.
3. Alternatively, there may be multiple genomes or multiple environmental samples for which reconstructed metabolism needs to be compared. In this case downloaded KO assignment files must be combined in a special way to allow comparison (*see* next section).

3.4 Further KEGG Mapper Analysis

The BlastKOALA result page contains direct links to the KEGG Mapper tools: Reconstruct Pathway, Reconstruct Brite and Reconstruct Module. This is equivalent to downloading the KO assignment file and uploading it for use by these tools at the KEGG Mapper website (www.kegg.jp/kegg/mapper.html). After the expiration of the result page or when the downloaded file is modified it is necessary to use KEGG Mapper this way. Here are some tips.

1. A file to be uploaded contains K numbers in the second column. This is different from the other KEGG Mapper tools, which take the first column as an identifier column.
2. When two or more datasets are to be compared, add a comment line starting with # at the beginning of each KO assignment file, and concatenate all files into a single file for use in Reconstruct Pathway and Reconstruct Brite tools. Note that global maps (map numbers 01100s) and overview maps (map numbers 01200s) allow comparison of only two datasets.
3. Reconstructed pathway maps and Brite hierarchies are distinguished by color coding as define in the KEGG color codes (www.kegg.jp/kegg/kegg1c.html). When multiple datasets are compared, regular maps are shown with split coloring for mapping multiple data to the same node (box), while global and overview maps are shown with a third color for mapping two data to the same edge (line).
4. Reconstruct Module tool can only be used for a single dataset. Instead, it has an option to include incomplete modules, for each module is checked for its completeness according to the logical expression. Incomplete, but almost complete modules may facilitate finding missed annotations.
5. Regular maps are represented by png image files, but global and overview maps are represented differently. Use the "Image (png) file" link to generate an image file. A publication-quality, high-resolution map can be obtained by generating a larger image file with a larger magnification and then changing the resolution, for example, from 72 dpi to 300 dpi.

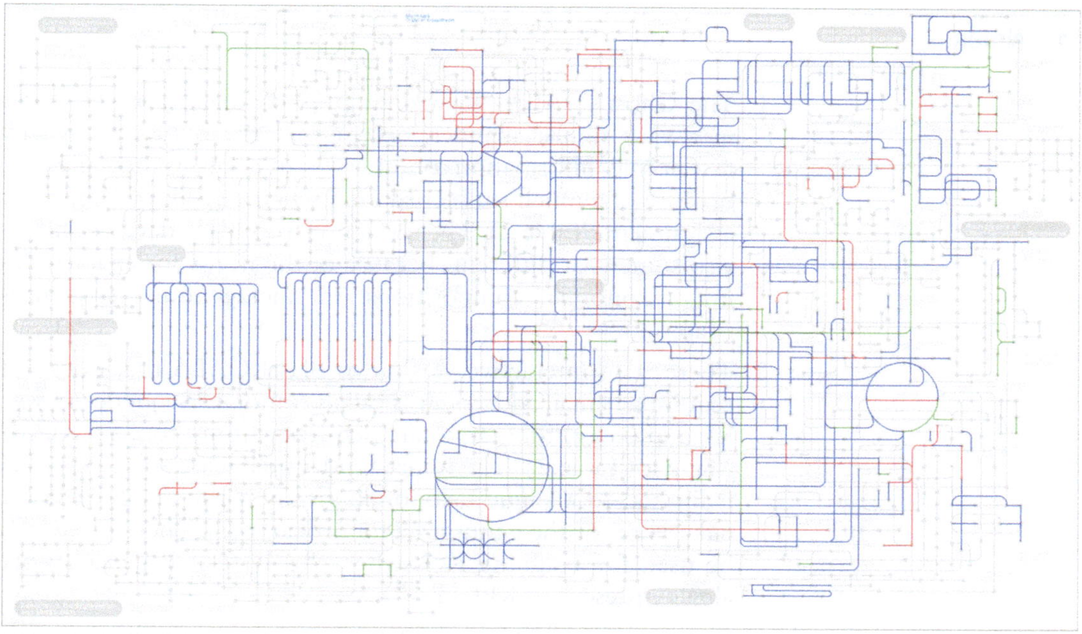

Fig. 2 A comparison of global metabolism in two organisms. Metabolic pathways in the KEGG global map (map01100) are shown with the following coloring: *green* for *Lentibacillus amyloliquefaciens* only, *red* for *Oceanobacillus iheyensis* only, and *blue* for both sharing

As an example, two genomes were compared by concatenating two KO assignment files with # lines. One was *Lentibacillus amyloliquefaciens* (GenBank accession: CP013862) [11] analyzed by BlastKOALA and the other was *Oceanobacillus iheyensis* (KEGG organism code: oih) [12] present in KEGG. Figure 2 shows the reconstructed global map of metabolic pathways (map01100), where the pathways present only in Lentibacillus or Oceanobacillus are shown in green or red and the pathways present in both are shown in blue. A more detailed picture of individual pathways may be examined from regular maps where enzymes shared by both are shown in split coloring of boxes. Brite hierarchies such as Enzymes (ko01000) are also multicolored when viewed from the link such as "Show ko01000." Furthermore, two sets of reconstructed module lists may be compared for differences. One notable feature in this example was the presence of catechol meta-cleavage module (M00569) only in Lentibacillus. In general, modules are more directly linked to functional meanings than pathways and easier to capture key features of metabolism.

4 Notes

1. The GhostKOALA (www.kegg.jp/ghostkoala/) server works in a very similar manner, except that the option to specify taxonomic information of query sequences is not present, because of a much simpler scoring scheme used for K number assignment (Table 2). The set of nonredundant database files is also somewhat different, because of the needs for characterizing taxonomic compositions in metagenome sequences (Table 1).
2. There is an interactive version of BlastKOALA that does not require an email address. It is called Annotate Sequence (www.kegg.jp/kegg/tool/annotate_sequence.html) tool in KEGG Mapper (Table 2). The nonredundant database to be searched is limited to a single taxonomic group at the family level for eukaryotes or the genus level for prokaryotes. Therefore, it can be used for assigning KOs when closely related genomes are already present in KEGG.
3. In the GhostKOALA result page, there is another pie chart showing taxonomic compositions of query sequences, together with links for analyzing taxonomic data.
4. The EC numbers may be useful for integrating BlastKOALA results with outside resources, but it is highly recommended to use K numbers rather than EC numbers within the KEGG resource. For a historical reason the EC numbers can be used as identifiers in two KEGG Mapper tools, Search Pathway and Search&Color Pathway, but again its use is discouraged.

Acknowledgment

This work was partially supported by the National Bioscience Database Center of the Japan Science and Technology Agency. The computational resource for developing and servicing KEGG is provided by the Bioinformatics Center, Institute for Chemical Research, Kyoto University.

References

1. Kanehisa M, Sato Y, Kawashima M, Furumichi M, Tanabe M (2016) KEGG as a reference resource for gene and protein annotation. Nucleic Acids Res 44:D457–D462
2. McDonald AG, Tipton KF (2014) Fifty-five years of enzyme classification: advances and difficulties. FEBS J 281:583–592
3. Ogata H, Goto S, Fujibuchi W, Kanehisa M (1998) Computation with the KEGG pathway database. Biosystems 47:119–128
4. Kanehisa M, Sato Y, Morishima K (2016) BlastKOALA and GhostKOALA: KEGG tools for functional characterization of genome and metagenome sequences. J Mol Biol 428:726–731

5. Li W, Godzik A (2006) Cd-hit: a fast program for clustering and comparing large sets of protein or nucleotide sequences. Bioinformatics 22:1658–1659
6. Kanehisa M, Goto S, Sato Y, Kawashima M, Furumichi M, Tanabe M (2014) Data, information, knowledge and principle: back to metabolism in KEGG. Nucleic Acids Res 42: D199–D205
7. Kanehisa M (2013) Chemical and genomic evolution of enzyme-catalyzed reaction networks. FEBS Lett 587:2731–2737
8. Pearson WR (1991) Searching protein sequence libraries: comparison of the sensitivity and selectivity of the Smith-Waterman and FASTA algorithms. Genomics 11:635–650
9. Altschul SF, Madden TL, Schäffer AA, Zhang J, Zhang Z, Miller W, Lipman DJ (1997) Gapped BLAST and PSI-BLAST: a new generation of protein database search programs. Nucleic Acids Res 25:3389–3402
10. Suzuki S, Kakuta M, Ishida T, Akiyama Y (2014) GHOSTX: an improved sequence homology search algorithm using a query suffix array and a database suffix array. PLoS One 9: e103833
11. Wang JL, Ma KD, Wang YW, Wang HM, Li YB, Zhou S, Chen XR, Kong DL, Guo X, He MX, Ruan ZY (2016) Lentibacillus amyloliquefaciens sp. nov., a halophilic bacterium isolated from saline sediment sample. Antonie Van Leeuwenhoek 109:171–178
12. Takami H, Takaki Y, Uchiyama I (2002) Genome sequence of Oceanobacillus iheyensis isolated from the Iheya Ridge and its unexpected adaptive capabilities to extreme environments. Nucleic Acids Res 30:3927–3935

Chapter 12

Ortholog Identification and Comparative Analysis of Microbial Genomes Using MBGD and RECOG

Ikuo Uchiyama

Abstract

Comparative genomics is becoming an essential approach for identification of genes associated with a specific function or phenotype. Here, we introduce the *m*icro*b*ial *g*enome *d*atabase for comparative analysis (MBGD), which is a comprehensive ortholog database among the microbial genomes available so far. MBGD contains several precomputed ortholog tables including the standard ortholog table covering the entire taxonomic range and taxon-specific ortholog tables for various major taxa. In addition, MBGD allows the users to create an ortholog table within any specified set of genomes through dynamic calculations. In particular, MBGD has a "My MBGD" mode where users can upload their original genome sequences and incorporate them into orthology analysis. The created ortholog table can serve as the basis for various comparative analyses. Here, we describe the use of MBGD and briefly explain how to utilize the orthology information during comparative genome analysis in combination with the stand-alone comparative genomics software RECOG, focusing on the application to comparison of closely related microbial genomes.

Keywords Orthology analysis, Clustering, Comparative genomics, Phylogenetic profile, Gene order comparison, Microbial genome

1 Introduction

A sequence similarity search using tools such as BLAST has been the most popular approach for assignment of function to a protein sequence. In particular, annotation transfer from the most significant hit is a common method of functional annotation for a newly determined genome. This "best hit strategy for annotation transfer" is linked to the concept of orthology. Orthology and paralogy are homologous relations derived from speciation and duplication, respectively [1]. Because a duplicate gene is generally free from selection and accumulates mutations rapidly, paralogs tend to have different functions, whereas orthologs tend to share equivalent functions. Therefore, orthologs rather than paralogs should be used to transfer annotation from different species. The best-hit

Daisuke Kihara (ed.), *Protein Function Prediction: Methods and Protocols*, Methods in Molecular Biology, vol. 1611, DOI 10.1007/978-1-4939-7015-5_12, © Springer Science+Business Media LLC 2017

strategy is effective at identification of orthologs because an ortholog of a given gene in a different species (if any) is always the closest homolog in that species (unless horizontal gene transfer is assumed).

Orthology is a key concept for comparative genome analysis, where orthology relations are used to establish correspondence between genes in different genomes. By means of the comparative genomics approach, several additional methods can be used for functional inference, including phylogenetic profile [2], domain fusion [3], and gene neighborhood methods [4]. Lately, the vast majority of protein sequences are being generated from high-throughput genomic sequence projects. Tens of millions of sequences are now registered in the UniProt protein sequence database, most of which come from complete or nearly complete proteome sets. Thus, the comparative genomics approach is becoming essential for analysis of newly determined genomic sequences by means of the existing sequence information through similarity searching. In particular, recent data accumulation allows researchers to compare genomes among closely related organisms that have different phenotypic traits; this approach provides valuable clues to the gene functions related to the phenotypic difference.

MBGD [5, 6] is a *m*icro*b*ial *g*enome *d*atabase for comparative analysis based on orthology analysis. It stores complete genome sequences of microbes (bacteria, archaea, and unicellular eukaryotes with some additional higher eukaryotes) determined so far as well as all-against-all similarity search results among the translated protein sequences identified in these genomes. The central function of MBGD is creation of ortholog groups within a specified set of genomes from these similarity relations using a specialized program, DomClust [7]. The grouping result can be represented as an ortholog table, with rows as ortholog groups and columns as genomes (organisms), thereby indicating the presence or absence of each ortholog in each genome. In contrast to many other methods for orthology inference that rely on bidirectional best-hit heuristics, DomClust is based on the traditional hierarchical clustering procedure using all-against-all similarities, which can naturally handle one-to-many or many-to-many orthologous relations [or in-paralogous relations, i.e., paralogs generated by duplication in one lineage after speciation [8]]. Moreover, DomClust can identify orthologs at the subgene level, i.e., a gene can be split into domains if needed due to a gene/domain fusion event. The accuracy of the domain level classification in DomClust is sometimes limited because it is based only on pairwise sequence alignment information. Therefore, MBGD also uses the DomRefine program [9] to improve ortholog grouping at the domain level on the basis of multiple sequence alignments.

In MBGD, several ortholog tables are precomputed, including the standard (default) ortholog table created from a representative set of genomes (containing one genome from each genus) covering the entire taxonomic range, and taxon-specific ortholog tables created for each major taxon in various taxonomic ranks (e.g., species, genus, and family). In addition, as a notable feature of MBGD in contrast to the existing "static" orthology databases, it allows the users to create an ortholog table dynamically by specifying an arbitrary set of genomes to compare. In particular, MBGD has the "My MBGD" mode [10] where users can upload their original genome sequences and incorporate them into the analysis through dynamical execution of similarity searching and ortholog clustering. Although MBGD uses only complete genome data for comparative analysis by default, many incomplete genome data have recently accumulated in the public database. MBGD now stores these draft genome data without precalculating all-against-all similarities [5]. In My MBGD mode, users can also utilize these draft genome data in orthology analysis through a dynamical similarity calculation.

Although MBGD provides several functions for retrieval and analysis of the resulting ortholog table for further comparative analysis, its analytical functionality is rather limited. Therefore, to complement MBGD, we also developed a stand-alone software application, named RECOG, which offers users several functionalities to visualize a large ortholog table more flexibly and thereby to conduct detailed comparative analyses. Although RECOG was originally designed as a client-server system comprising a set of programs to construct a data analysis server (RECOG server) locally that is almost equivalent to MBGD, its installation and setup are rather laborious. Therefore, here, we mainly describe the use of MBGD and briefly explain how RECOG can be used to visualize the ortholog tables created in MBGD.

2 Materials

2.1 Input Data

If the sequences to be compared are only the genome data in the public database, you need not prepare anything. If you have your own genome data to be analyzed, you will need some preparation. First, the coding sequences (CDS) should be identified in the genome sequences because MBGD compares only sets of protein sequences. Prepare the translated sequences in FASTA format and a gene location table as a tab-delimited file, in addition to the genomic DNA sequence. *See* Subheading 3.4 for details. Alternatively, all of these data can be combined in a GenBank format file.

There are three types of input data: For complete genome data, you should prepare the data for each chromosome/plasmid. If your data are draft genome data containing multiple contigs, you can put

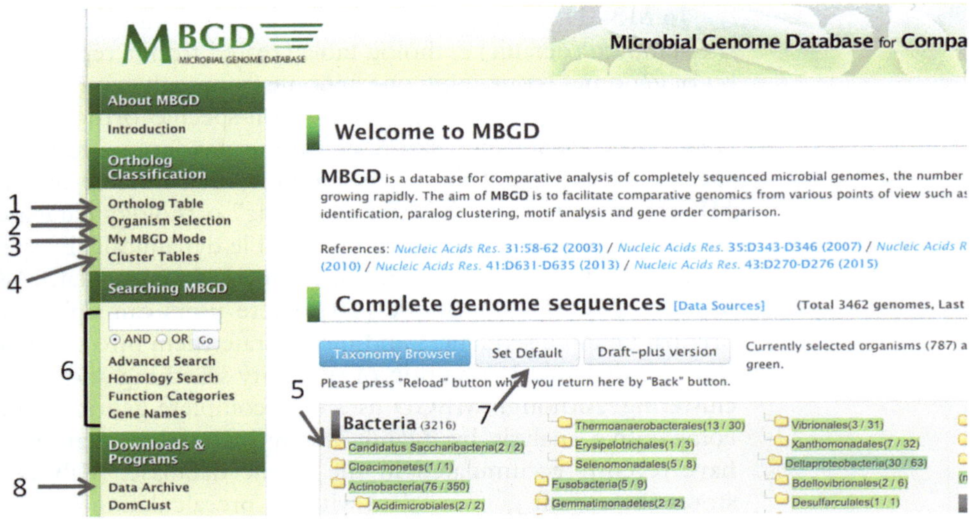

Fig. 1 The top page of MBGD. (*1*) Browse the current ortholog table or switch to another ortholog table; (*2*) Create a new ortholog table using a specified set of genomes and parameters; (*3*) Go to My MBGD mode and upload user genomes; (*4*) Show a list of ortholog tables accessed so far; (*5*) Show an individual genome map; (*6*) Keyword search and other search functionalities; (*7*) Set the default ortholog table as the current ortholog table. (*8*) Download the archives of the current/previous versions of MBGD

the data of multiple contigs into one file. Alternatively, you can prepare only a set of protein sequences in FASTA format. The last type is effective for low-quality draft genome data or transcriptome data, although some functionality such as genome map comparison is not available for this type of data.

2.2 Software

To use RECOG, download the RECOG client program for your operating system from http://mbgd.genome.ad.jp/RECOG/ and install it. Because RECOG is implemented in Java, the Java runtime environment should also be installed. Optionally, you can download and install the RECOG server program from that web page if you want to construct the database locally although here we do not mention this approach further.

2.3 The Web Site

Access the MBGD web site at http://mbgd.genome.ad.jp/ (Fig. 1).

3 Methods

3.1 Overview of MBGD

As explained in Subheading 1, MBGD retains many ortholog tables, including precomputed ones and dynamically created ones. Each ortholog table is associated with a set of parameters, including the target set of genomes to compare and the parameters of the clustering program (DomClust). For each session, MBGD designates one ortholog table as a "current ortholog table" and uses it for comparative analysis. By default, the current ortholog table is the standard ortholog table, which is created from the set of genomes containing one representative genome from each genus covering the entire taxonomic range.

To browse the current ortholog table or to switch it to another existing ortholog table, click "Ortholog Table" on the top page of MBGD (Fig. 1, arrow 1; *see* Subheading 3.2). To create a new ortholog table by choosing a set of genomes to compare from the existing genomes and/or changing the clustering parameters, click "Organism Selection" (Fig. 1, arrow 2; *see* Subheading 3.3). If you have some original genome data that you want to analyze, you can add them to MBGD through "My MBGD" mode (Fig. 1, arrow 3; *see* Subheading 3.4). In this mode, you can also use the published draft genome data that are stored in MBGD but are not included in the all-against-all similarity precomputation. You can see the list of ortholog tables that you have created/used and the set of parameters associated therewith by clicking "Cluster Tables" (Fig. 1, arrow 4), where you can choose a table by clicking a unique ID assigned to each ortholog table.

MBGD provides several search functionalities including gene search on a genome map (Fig. 1, arrow 5), quick keyword search, advanced keyword search, and homology search with the user's query sequence (Fig. 1, area 6). By means of these functions, the users can first search for a gene of particular interest and then retrieve the ortholog group that the gene belongs to in the current ortholog table. Once the current ortholog table has been changed, the target genomes for these search functions are confined to those contained in the current ortholog table. The current target genomes are highlighted within the taxonomy box on the top page (Fig. 1, arrow 5). The users can reset the current ortholog table to the default one, by clicking the "Set Default" button (Fig. 1, arrow 7). Finally, if you need the entire data in the current or previous MBGD releases, they can be downloaded from the Data Archive page (Fig. 1, arrow 8).

3.2 Viewing Precomputed Ortholog Tables

1. On the MBGD top page, click Ortholog Table (Fig. 1, arrow 1).
2. The Ortholog table overview page is displayed. The standard (default) ortholog table is created from the representative genome set containing one genome from each genus. The users can choose a taxon-specific ortholog table from the left-hand taxonomy tree (Fig. 2a, arrow 1) to change the current ortholog table.
3. On this page, the users can change the view by selecting a tab at the top of the view (Fig. 2a, arrow 2). The available views include:

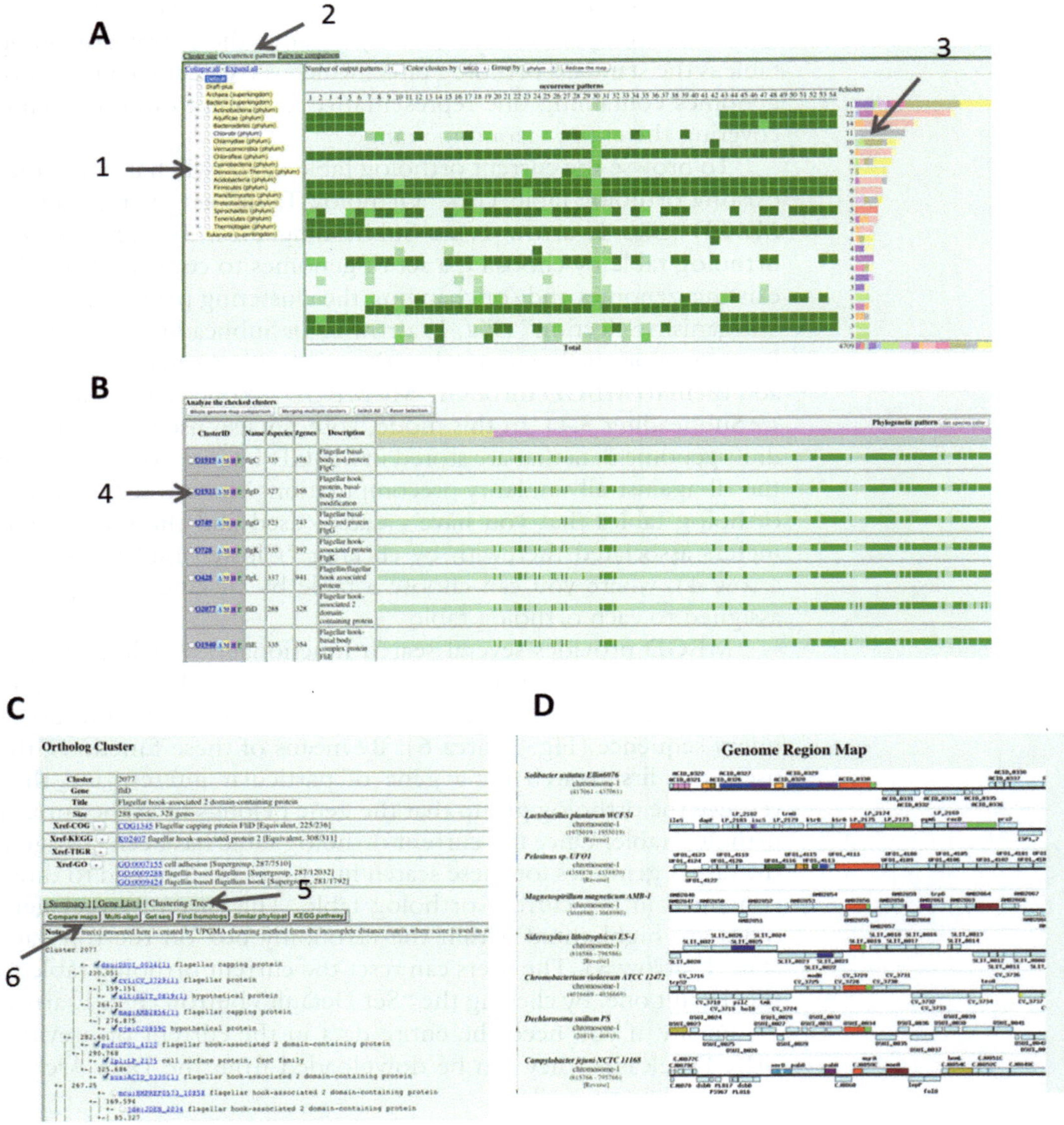

Fig. 2 Browsing an ortholog table in MBGD. (**a**) The ortholog table overview page. (*1*) The list of ortholog tables including the standard (default) ortholog table, taxon-specific ortholog tables, and user-generated ortholog tables; (*2*) Tabs for switching the view; (*3*) A bar graph showing the number of ortholog groups representing each occurrence pattern. (**b**) Ortholog table viewer. (*4*) The ortholog ID in the first column is a link to the ortholog cluster information page. (**c**) The ortholog cluster information page. (*5*) Tabs for switching the display for the cluster members; (*6*) Buttons of analytical functions. (**d**) A comparative genome region map

- *Cluster size*: the default view showing the histogram of cluster size.
- *Occurrence pattern*: a bar graph showing the relation between occurrence patterns (representing presence or absence in each genome) and functional categories.

- *Pairwise comparison*: a similarity matrix for a pairwise genome comparison. Two types of matrix can be shown: (1) the ratio of the shared orthologs or (2) the average percent identities among orthologs, within each pair of genomes. Dot plot analysis within the specified pair of genomes can also be conducted using the CGAT interface [11] by clicking the corresponding cell of the matrix.
- *CoreAligner*: a diagram showing the syntenically conserved core structure identified by the CoreAligner program [12]. This function is effective only for comparisons of closely related genomes, and thus is available only for a family or lower-level taxon.

Here, let us choose the occurrence pattern view (Fig. 2a). By clicking an appropriate bar in this bar graph (Fig. 2a, arrow 3), the corresponding subset of the ortholog table is displayed (Fig. 2b).

4. In the ortholog table viewer (Fig. 2b), the occurrence pattern for each ortholog group is displayed in each row. After a click on the link on an ortholog group ID in the header column (Fig. 2b, arrow 4), detailed information about the ortholog group is displayed (Fig. 2c).
5. On the ortholog cluster information page (Fig. 2c), several pieces of information about the given ortholog group are displayed, including the consensus annotation (gene name and title) and cross-references to other ortholog databases (COG [13], KEGG [14], and TIGRFAMs [15]) and the GeneOntology database [16], in addition to the members of this ortholog group shown in the lower box. The view within this box can be changed by selecting the tabs (Fig. 2c, arrow 5): after a click on the "Gene List" tab, genes belonging to this ortholog group are tabulated for each organism grouped according to the taxonomic classification, and after you select the "Clustering Tree" tab, the hierarchical clustering tree calculated by DomClust/DomRefine is displayed (Fig. 2c).
   On this page, several analysis functions can be invoked by pressing the buttons below these tabs (Fig. 2c, arrow 6), including
   - *Compare maps*: display multiple genome maps in which the genes of the target ortholog group are aligned at the center (colored red) and the genes belonging to the same ortholog group are assigned the same color and pattern (Fig. 2d). This way, the users can compare the gene orders around the orthologs in different genomes (*see* **Note 1**).
   - *Multi-align*: calculate the multiple sequence alignment of the genes in this ortholog group (as protein or DNA sequences) using Clustal Omega [17] or MAFFT [18],

and create a phylogenetic tree from the resulting alignment using FastTree [19] (*see* **Note 1**).

- *Find homologs*: find ortholog groups that are homologous to this ortholog group. The result will be displayed in the ortholog table viewer (*see* **Note 2**).
- *Similar phylopat*: search for ortholog groups that have an occurrence pattern similar to that of this ortholog group. A dissimilarity measure should be chosen from the following: a correlation coefficient, hamming distance, and mutual information (*see* **Note 3**). The result will be displayed in the ortholog table viewer as an ordered list according to the calculated dissimilarity.

3.3 Creating an Ortholog Table Using a Selected Set of Genomes

In addition to the precomputed ortholog tables, the users can dynamically create an ortholog table by choosing an arbitrary set of genomes.

1. On the top page of MBGD, click "Organism Selection" (Fig. 1, arrow 2).
2. The taxonomy browser is displayed. Press the "Clear All" button to reset the current selection (if you want). Then, choose a set of genomes to compare by clicking the checkboxes (*see* **Note 4**). After the selection is completed, press the "Choose checked taxa" button.
3. Here, you can change the clustering parameters, if required, by pressing the "Change Clustering Parameters" button (*see* **Note 5**). Next, press the "Create/View Ortholog Cluster" button to start the clustering process.
4. After the execution is completed, the ortholog cluster overview page (Fig. 2a) is displayed. *See* Subheading 3.2 regarding subsequent analysis.

3.4 Ortholog Analysis with User Genome Sequences and/or Draft Genome Sequences in My MBGD Mode

1. On the top page of MBGD, click "My MBGD Mode" (Fig. 1, arrow 3). In this mode, you can upload genome data and add them to comparative analysis.
2. The Data input page is displayed (Fig. 3a; *see* **Note 6**). If you want to analyze your original genome data, you can add them here. For this purpose, first, enter the information about the genome, including species name, abbreviated species code (*see* **Note 7**), NCBI Taxonomy ID (*see* **Note 8**), and data type (complete genome, draft genome, or proteome; *see* Subheading 2.1). Then press the "Save" button. Alternatively, if the genome data you want to add are only draft genome data already stored in MBGD, then press the "Next" button to skip the current step and go to **step 4**.

A

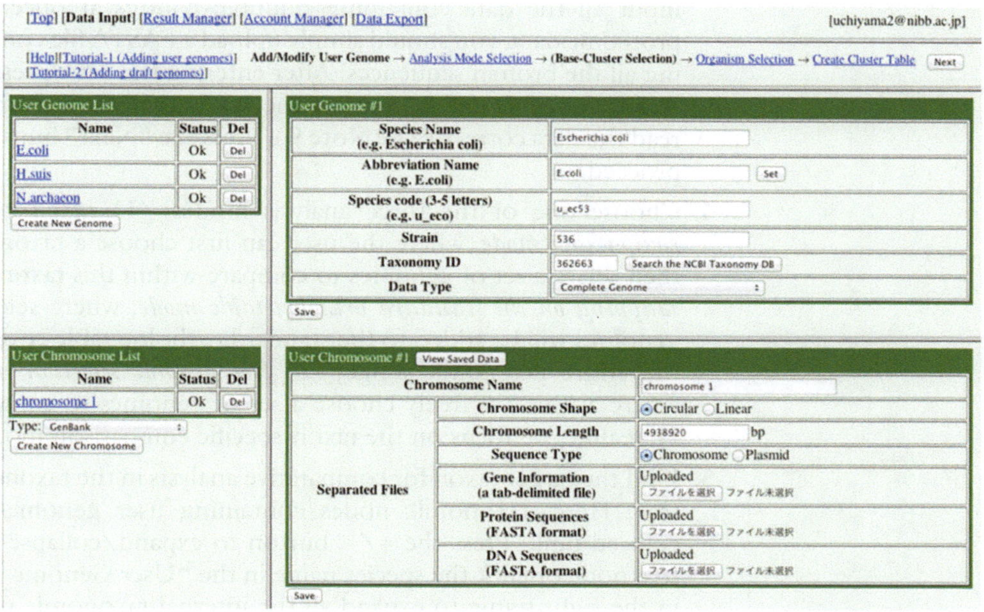

B

RBE_0001		1	822	1	hypothetical protein
RBE_0002	trxA	849	1208	1	Thioredoxin
RBE_0003	rfbE	1220	1966	1	O-antigen export system ATP-binding protein Rf
RBE_0004	rfbA	1963	2739	1	O-antigen export system permease protein RfbA
RBE_0005	gltD	2736	5990	1	NADPH-dependent glutamate synthase beta cha
RBE_0006	lexA	6013	6432	-1	SOS-response transcriptional repressors (RecA-
RBE_0007		6499	7605	-1	Acyltransferase family protein
RBE_0008		7636	8670	-1	hypothetical protein

C

#id	name	chrid	from	to	dir	type	descr
I738_RS0100010		NZ_AQYW01000001	420	848	-1	CDS	hypothetical protein
I738_RS0100015		NZ_AQYW01000001	913	1092	-1	CDS	hypothetical protein
I738_RS0100025		NZ_AQYW01000002	409	1026	-1	CDS	hypothetical protein
I738_RS0100035		NZ_AQYW01000003	8	781	1	CDS	hypothetical protein
I738_RS0100045		NZ_AQYW01000004	25	1245	-1	CDS	hypothetical protein
I738_RS0100050		NZ_AQYW01000005	75	1289	1	CDS	hypothetical protein
I738_RS0100055		NZ_AQYW01000005	1296	1544	-1	CDS	30S ribosomal protein S11

Fig. 3 User genome input in My MBGD mode. (**a**) The data input form in My MBGD mode. (**b**) The simple format that consists of the following six columns: gene ID, gene name, start position, end position, direction, and description of protein function. (**c**) The extended format in which the first line beginning with the hash character (#) indicates the type of data in each column. The latter format includes the Chromosome ID (chrid) column and thereby allows the users to place the information on multiple chromosomes or contigs into one file. It also includes the Gene type (type) column to specify the location of noncoding genes (for drawing a genome map), although MBGD uses only coding sequences (CDS) in orthology analysis

3. Enter the sequence data for each chromosome as either a GenBank format file or a combination of a protein sequence FASTA file, a DNA sequence FASTA file, and a tab-delimited gene location table. The gene location table should contain the following columns in this order: gene ID (locus tag), gene name, start position, end position, direction (1/−1), and description of protein function (Fig. 3b). Alternatively, you can specify the data type of each column in the first comment line beginning with a hash character (Fig. 3c). Here, the fields of the input form depend on the data type specified above. For complete genome data, you should enter the data for each

chromosome/plasmid. For incomplete genome data, you can input all the data containing multiple contigs at once. For proteome data, you should simply upload a FASTA file containing all the protein sequences. After entering the data, press the "View Saved Data" button to check whether the server can read the data correctly (*see* **Note 9**). Press the "Next" button to proceed.

4. Choose one of the three analysis modes: (1) *taxon-specific comparison mode*, where the user can first choose a taxon and then select a set of genomes to compare within this taxon; (2) *mapping on the standard ortholog table mode*, where selected genomes will be added to the standard ortholog table covering the entire taxonomic range; (3) *free genome selection mode*, where a user can freely choose a set of genomes to compare. Hereafter, we focus on the taxon-specific comparison mode.
5. Find the target taxon for comparative analysis in the taxonomic tree. Here, taxonomic nodes containing user genomes are colored pink. Press the +/− button to expand/collapse each tree node or click the species name in the "User Genome List" in the right frame to expand all the internal taxonomic nodes containing the target genome. Click the radio button to choose the target taxon and press the "Next" button.
6. Genomes belonging to the specified taxon are listed in the tree. If you want to add draft genomes, press the "Show Draft (Reload)" button to show the draft genomes. By default, the representative genomes used for precalculation of the taxon-specific ortholog table of this taxon (if available) are checked, and the users can add user genomes to this precalculated ortholog table (called *base cluster*) using an incremental clustering procedure (MergeTree). If you want to create the clustering from scratch (i.e., not to use the base cluster), press the "Clear All" button to cancel this selection. Here, draft genomes and user genomes are colored pink and light blue, respectively. In My MBGD mode, you should specify at least one draft genome or user genome (otherwise, you should follow the procedure in Subheading 3.3). You can also use the checkbox in the leftmost column of the "User Genome List" in the right frame to choose user genomes. After specifying the set of genomes to compare, press the "Next" button.
7. Check the list of selected species and press the "Create/View Cluster Table" button to start execution. First, sequence similarities between query genomes (user genomes and draft genomes) and complete genomes stored in MBGD are calculated using the same protocol for constructing MBGD (i.e., using BLASTP [20] followed by Smith-Waterman alignment [21]). After the resulting similarity relations are stored in the database, ortholog clustering by DomClust starts.

8. After the execution is completed, the ortholog table overview page (Fig. 2a) is displayed. *See* Subheading 3.2 regarding subsequent analysis.
9. After the analysis is finished, you should exit My MBGD mode by pressing "Exit My MBGD Mode" at the top page of MBGD before using MBGD with the default ortholog table or other ortholog tables that do not contain user-specific genomes (*see* **Note 10**).

3.5 Visualizing the Ortholog Cluster Table Using RECOG

RECOG is a stand-alone software tool for comparative genome analysis, which allows users to browse and manipulate a large ortholog table like an Excel table. Although RECOG was originally designed for use via a connection to a specialized RECOG server, it can also import and display the clustering output created by the MBGD server and thus enables the user to conduct a more thorough comparative analysis (although some functionalities are not available without connecting to a valid RECOG server). Here, we briefly introduce RECOG through this kind of use.

1. First, you should download the ortholog clustering data that you want to analyze from MBGD. For this purpose, go to the ortholog table overview page (Fig. 2a), choose the ortholog table that you want to analyze, and find the button "Save Complete Table" at the bottom of the "Search conditions" form. Choose "dclst" format and save the data by pressing the "Save Complete Table" button. You should also save the taxonomy tree containing the current genome set by pressing the "Save Taxonomy Tree" button. The clustering data and taxonomy data will be saved as files named "cluster.dclst" and "cluster.tax," respectively, but you can rename them if you like.
2. Launch the RECOG client program. First, you should create a project, which is a folder in which all the files will be created during the subsequent analysis. To create a project, select "File > New > New Project" from the menu bar and name the project.
3. Import the clustering output file by selecting "File > Import > DomClust file" in the menu bar and choose the above clustering file from the file list. Also import the taxonomy tree by selecting "File > Import > Taxonomy tree" and choose the above-mentioned taxonomy tree file. Imported data is retained and does not need to be imported again; it can be loaded by simply clicking the folder icon or selecting "File > Open Data" and choosing the data from the file list (*see* **Note 11**).
4. The main window of the RECOG client program consists of three parts (Fig. 4). The central part shows the ortholog cluster table, in which each cell contains genes of a given ortholog

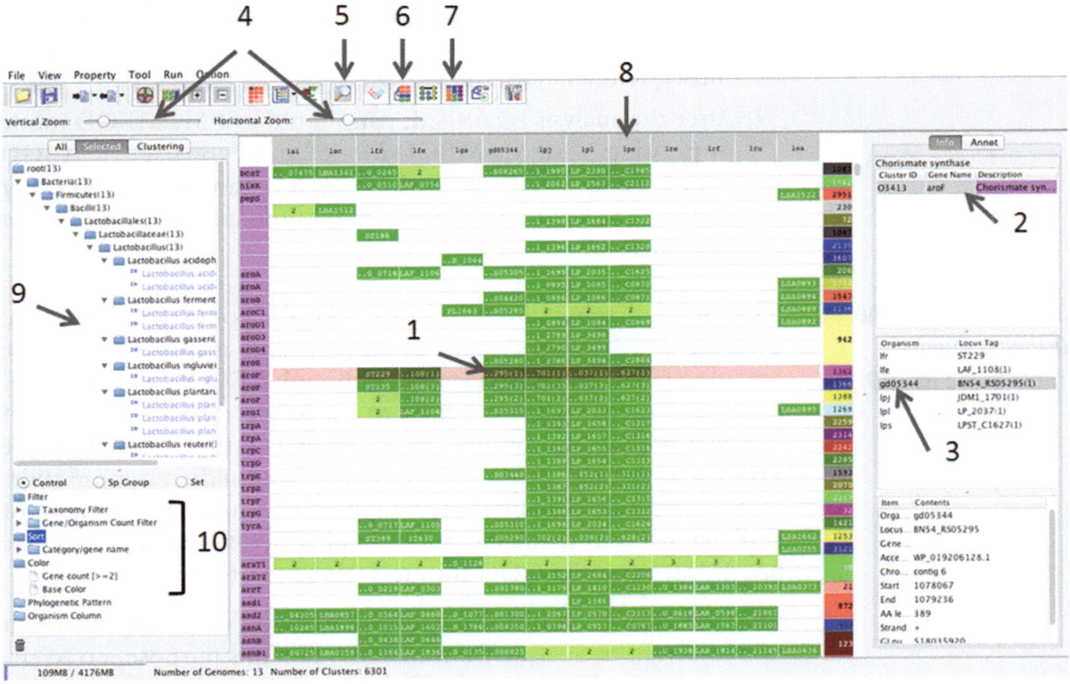

Fig. 4 The RECOG main window. (*1*) A cell of the ortholog cluster table is clicked. (*2*) The list of the selected ortholog clusters. (*3*) The list of genes included in the selected ortholog clusters. (*4*) Vertical and horizontal zoom sliders. (*5*) Keyword search. (*6*) Phylogenetic pattern clustering. (*7*) Neighboring clustering. (*8*) The header of the cluster table. (*9*) Taxonomy tree browser. (*10*) Control panel

group (row) in a given genome (column). If you click a cell in this table (Fig. 4, arrow 1), information about the clicked ortholog group is displayed in the right-hand panels, including ortholog group ID and annotation (upper right), members of this ortholog group (middle right), and information on the selected gene (lower right). By double-clicking an ortholog ID (Fig. 4, arrow 2) or a gene name (Fig. 4, arrow 3), you can access the MBGD database to get detailed information about that ortholog or gene, respectively. Alternatively, the annotation of each ortholog group can be displayed in each row in the right-hand panel if you select the "Annot" tab. The size of each cell can be changed by means of the vertical and horizontal zoom sliders above the table (Fig. 4, arrow 4).

5. RECOG offers several functions for the analysis of the ortholog table. Each analytical function modifies the ortholog table by changing the following three elements of the table view: *filtering* (displaying only selected rows), *sorting* (ordering rows), and *coloring* (coloring genes/cells/rows). The most common functions are accessible via the icons listed at the top of the window, including:

- *Keyword search* (Fig. 4 arrow 5): search for keywords and display the search results in the ortholog table by filtering out unmatched ortholog groups and/or coloring matched genes
- *Phylogenetic pattern clustering* (Fig. 4, arrow 6): perform hierarchical clustering of ortholog groups based on phylogenetic pattern (occurrence pattern) similarity and sort the cluster table according to the clustering results. The hierarchical clustering tree is shown in the left-hand panel.
- *Neighboring clustering* (Fig. 4, arrow 7): perform clustering of genes in the same genome that are closely located in the chromosomal sequence as well as in the ortholog table, and assign the same color to the genes belonging to the same cluster. This function helps to identify syntenically conserved genes (as genes with the same color in the same column) by working in tandem with the sorting table function based on the chromosomal order in a particular genome, which can be invoked by double-clicking the header line of the target genome (Fig. 4, arrow 8).

6. In the taxonomy tree shown in the upper left panel (Fig. 4, arrow 9), the users can choose a set of genomes and apply several operations to the selected genomes from the menu shown after right-clicking. In particular, the function called *Taxonomy filtering* gives users a flexible way to specify the occurrence pattern. Here, the users can specify the following conditions for each selected genome set: *All* (present in all the specified genomes), *Any* (present in a subset of the specified genomes), and *None* (absent); for *Any* condition, the users can specify more detailed conditions (*see* Subheading 5 and Fig. 5c therein)
7. In the lower left panel, the users can check and modify the current status of the cluster table view in terms of three elements: Filter, Sort, and Color (Fig. 4, area 10). Here, the effects of the elements currently applied are listed under the tree and the users can inactivate/reactivate each effect by double-clicking it or can modify its setting by right-clicking it to bring up the relevant menu (*see* **Note 12**). In addition, the users can modify the order and visibility of columns (genomes) in the "Organism Column" section.

4 Case Study

In the text below, we will show a sample session for the comparative analysis of *Lactobacillus* species associated with weight gain and weight protection, according to ref. [24]. Here, the following

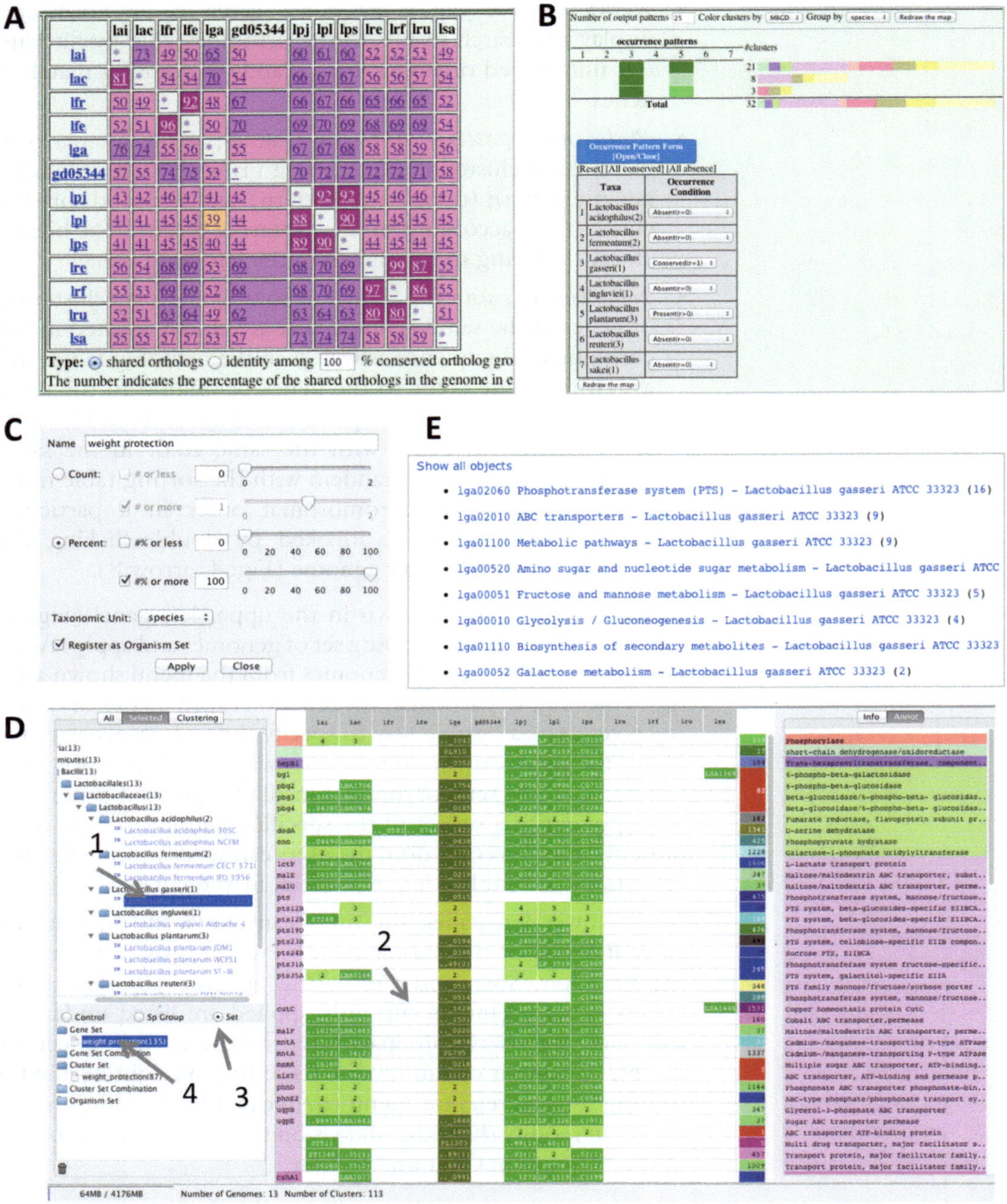

Fig. 5 A sample session for comparative analysis of *Lactobacillus* genomes. (**a**) A distance matrix showing the ratio of the shared orthologs among the *Lactobacillus* genomes. (**b**) The occurrence pattern view showing the weight protection-specific occurrence pattern. (**c**) The dialog box where you can specify detailed conditions for the *Any* taxonomy filter. (**d**) The result of the taxonomy filtering. *Arrows 1* through *4* indicate the steps for saving the genes in *L. gasseri* included in this filtered cluster set as a gene set. (**e**) The result of the KEGG pathway search. Click "Show all objects" to see the detailed gene list included in each pathway

4.1 Comparative Genomics of Lactobacillus Species Associated with Weight Gain and Weight Protection

Lactobacillus genomes (13 strains in seven species) are compared: *L. acidophilus* 30SC and NCFM (two strains); *L. fermentum* CECT 5716 and IFO 3956 (two strains); *L. ingluviei* Autruche 4; *L. reuteri* DSM20116, JCM1112, and SD2112 (three strains); *L. sakei* 23K; *L. gasseri* ATCC 33323; *L. plantarum* JDM1, ST-III, and WCFS1 (three strains). Among them, the last two species (*L. gasseri* and *L. plantarum*) are associated with weight loss (weight protection) and the remaining five species are associated with weight gain [24], and the purpose here is to identify candidate genes contributing to the differences between these two sets of species, especially those involved in carbohydrate and lipid metabolism that can be associated with weight modifications [24].

Among these genomes, *L. ingluviei* Autruche 4 is a draft genome that is not included in the MBGD complete genome set. Therefore, we should use My MBGD mode (Subheading 3.4) to incorporate it. Here, we do not have original genome data and thus skip the Data Input page (Fig. 3a) shown after entering My MBGD mode. Choose the taxon-specific comparison mode, and then find and pick the genus *Lactobacillus* by traversing the taxonomy tree as follows: Bacteria (superkingdom) > Firmicutes (phylum) > Bacilli (order) > Lactobacillales (family) > *Lactobacillus* (genus). Press the "Show Draft (Reload)" button to display draft genomes, press "Clear All," and then choose the 13 genomes listed above. Then, press the "Next" button and start execution.

After the execution is finished, you can see the histogram of cluster sizes. First, select the "Pairwise comparison" tab to show the similarity matrix to confirm the genomic diversity between the compared strains. In this matrix, you can see that the ratios of the shared orthologs between different species are generally 40–70%, whereas those within the same species are generally 80–99% (Fig. 5a), indicating substantial diversity of gene content between different *Lactobacillus* species. You can also check the average amino acid percent identities of orthologs by clicking the identity radio button, and see that the identities between different species are generally 50–70%, whereas those within the same species are 88–99%.

Next, select the "Occurrence pattern" tab to conduct occurrence pattern analysis (Fig. 5b). Here, a condition in the occurrence pattern can be specified in the Occurrence Pattern Form below the bar graph. Choose "Conserved ($r = 1$)" for *L. gasseri*, "Present ($r > 0$)" for *L. plantarum*, and "Absent ($r = 0$)" for other species to specify the "weight protection species-specific" occurrence pattern and press the "Redraw the map" button beneath the form, then 32 ortholog groups are found in total (Fig. 5b). To see the list of the resulting genes, click the "Total" bar near its left end. In this list, you can easily see that it contains many of the genes involved in phosphotransferase system (PTS), a sugar uptake system driven by phosphoryl group transfer.

To confirm this observation, let us try to use the link to the KEGG database. First, press the "Select All" button within the "Analyze the checked clusters" box above the table to check all the ortholog groups, and press the "Merging multiple clusters" button in the same box. The result is an ortholog cluster page containing all the selected ortholog groups. Press the "KEGG pathway" button and choose the organism *L. gasseri*, then press the "Exec" button. After that, you can jump to the KEGG Pathway database and see the list of *L. gasseri* pathways containing the genes included in these ortholog groups. On this page, you can confirm that PTS is the pathway that contains the largest number of "weight protection species-specific" genes.

Next, to analyze these data further using RECOG, save the table in the "dclst" format and rename the saved file as "lactobacillus.dclst." You should also save the taxonomy tree and rename it "lactobacillus.tax." Then, launch the RECOG client program. Create a new project named "lactobacillus." Import the clustering data and taxonomy tree from the above-mentioned saved files (*see* Subheading 3.5). Select the "Annot" tab in the left-hand panel to display the annotation of each ortholog group.

In RECOG, an occurrence pattern can be specified more flexibly than in MBGD. In the taxonomy tree in the upper left panel, choose the set of "weight protection" species, i.e., *L. gasseri* and three *L. plantarum* strains and right click > Taxonomy Filter > Any New > Setting, then in the displayed dialog, enter the name as "weight protection," choose "Percent," specify the condition as "100%," choose "Species" for Taxonomic Unit, and place a checkmark next to "Register as Organism Set" (Fig. 5c). This procedure specifies the condition that orthologs should be present in at least one strain of both "weight protection" species. Next, choose the remaining nine genomes of the "weight gain" species, right-click them to display the above dialog, and specify the name as "weight gain," choose "Count," the condition as "1 or less," and "Species" for Taxonomic Unit. This procedure specifies the condition that orthologs should be present in at most one species among the "weight gain" species. This way, you can specify an occurrence pattern based on phenotypic groups rather than taxonomic groups and then search for the pattern allowing for some exception. This relaxed pattern search enables one to obtain more candidate genes specifically for the weight protection species (Fig. 5d).

For further analysis, you may want to export the set of genes included in the filtered ortholog table for use in external software. Now, we will demonstrate the procedure to find pathways that include the genes listed here using the KEGG pathway database. For this purpose, first, you should create a gene set that contains genes of a particular organism (here we consider *L. gasseri* again) included in this table as follows: Choose *L. gasseri* (Fig. 5d, arrow 1) in the left-hand taxonomy browser; click the central cluster table window (Fig. 5d, arrow 2) and press Ctrl + A (for Windows PC) or Command + A (for Mac) to select all the rows in this table; right-click on the same window and choose "Create gene set"; then, enter the name "weight protection specific" and press "Apply" in the dialog box to create a gene set. The created gene set appears in the lower left panel if you click the "Set" radio button (Fig. 5d, arrow 3) and double-click to open the "Gene Set" node (Fig. 5d, arrow 4). Right-click this gene set and choose "Export gene set" to save the data into a file with a gene set file format. The created file contains the names (locus tags) of the target genes and is used for the KEGG Search Pathway function (http://www.genome.jp/kegg/tool/map_pathway1.html) to search for pathways containing the target genes. For this purpose, you should specify in the "Search against" field the three-letter code of the target organism, which appears in the header row in the ortholog table (in this case, "lga" for *L. gasseri* ATTC33323). As a result of this search, you can see that the gene set contains a more comprehensive set of PTS genes, in addition to carbohydrate transporter genes and several enzymes involved in carbohydrate metabolism (Fig. 5e).

5 Notes

1. Displaying a comparative map or a multiple sequence alignment of a very large ortholog group can be very time consuming (a multiple sequence alignment is calculated dynamically except for the default ortholog table, in which case, the precomputed alignment is displayed) and can produce too large an output for a detailed comparison. Instead of using all the sequences in a given ortholog group, you can choose a subset of sequences to analyze. For this purpose, select the "Gene List" or "Clustering Tree" tab and choose the sequences that you want to compare using the checkboxes before starting the analysis. In the Gene List view, you can choose representative genes using the taxonomic classification, whereas in the Clustering Tree view, you can choose them using the hierarchical clustering information.
2. In MBGD, presence-absence analysis is based on ortholog grouping. This is reasonable because orthologs are better candidates for functionally equivalent genes than paralogs. Nonetheless, a lack of orthologs of a given gene in some organism does not necessarily mean that the organism is lacking the given gene function, because (1) ortholog classification may not be correct or (2) the function can be complemented by other more distantly related homologs (paralogs). The *Find homologs* function can be used to check these possibilities. During clustering, DomClust first creates homologous sequence groups

and splits each of them into multiple ortholog groups, then outputs the homologous group information as a higher-level group comprising multiple ortholog groups. The *Find homologs* function uses this information to identify the ortholog groups homologous to the given ortholog group. To check the above possibilities, you should compare genes in these homologous ortholog groups. To this end, you can merge multiple ortholog groups by first choosing ortholog groups using the checkbox in the leftmost column of the cluster table and then by pressing the "Merging multiple clusters" button. The resulting ortholog cluster page contains all the selected ortholog groups, from which you can compare genes belonging to different ortholog groups using the comparative map and multiple alignment functionalities.

3. These measures are defined according to ref. [22] but are normalized to serve as dissimilarity measures ranging between 0 (most similar) and 1 (most dissimilar). To be precise, the "correlation coefficient," d_r, is defined as $d_r = (1 - r)/2$ using the actual correlation coefficient r; "hamming distance," d_H, is defined as $d_H = h/N$, using the actual hamming distance h and the number of genomes N; and "mutual information," d_I, is defined as $d_I = 1 - I$ using the actual mutual information I. The default measure is d_r. In contrast, d_H is a more straightforward measure reflecting the string difference, but it underestimates the dissimilarity when the numbers of 0's (absence) or 1's (presence) in the compared patterns are extremely biased. The measure d_I is also not intuitive because the dissimilarity between the same patterns is not always 0. Nonetheless, this measure can be useful in that it equally evaluates both negative and positive correlation patterns. Genes of negative correlation (or complementary) occurrence patterns can have equivalent functions because such patterns may have arisen during evolution due to displacement of genes of a particular function with functionally equivalent nonorthologous genes (nonorthologous gene displacement) [23].
4. By clicking a higher-level taxon, you can select all the genomes belonging to the specified group. If you want to choose only representative species/strains from each taxon (e.g., species, genus), go to the "Choose one genome for each" section at the top of the page and select a taxonomic rank from which you want to choose a representative, and then press the "Select" button. The selection of representatives is based on the weight assigned to each organism. In MBGD, each genome is weighted according to the release date such that a genome published earlier has a greater weight.
5. There are several parameters to control the clustering output. Generally, modifying the parameters changes the granularity of

clustering; for example, using a more stringent cutoff produces more fine-grained clustering, i.e., a larger number of small clusters. The default parameter set was determined using the well-curated COG database as a reference [7]. Although it can also work for most other cases, you may want to customize it to your situation. Below are a few important points:

(a) The default similarity cutoff is rather loose. This is because there is no general lower limit for similarities among orthologs; nonetheless, the best-hit-based criteria can identify correct orthologs in most cases. If you want to eliminate possible false positives that have only weak similarity, you can use a lower BLAST E-value cutoff. Alternatively, if you are comparing genomes of the same or very closely related species, you may want to set a higher percent identity cutoff, e.g., 90%.

(b) As mentioned in **Note 2**, DomClust splits the clustering tree of homologous sequences to create ortholog groups. The criterion whether the tree should be split at a given internal node is $|A \cap B| / \min(|A|, |B|) > p$, where A and B are sets of species included in the subclusters of both sides of the node and p is the "Cutoff ratio of paralogs for tree splitting" parameter. The default setting is $p = 0.5$. Because p should be 0 in the rigorous orthology definition, a reduction in this parameter applies more rigorous orthology definition resulting in more fine-grained ortholog clusters. Nonetheless, using too rigorous a setting is generally not recommended because it can make the analysis more susceptible to errors due to a wrong gene tree topology or existence of horizontal gene transfer.

(c) One of the salient features of the DomClust algorithm is that it can split genes into domains, but the domain-splitting procedure gives rise to another delicate problem: over-splitting. To reduce this problem, the following parameters can be adjusted to control the extent to which genes will be split: "Alignment coverage for domain splitting," "Score cutoff for domain splitting," "Overlap ratio for merging adjacent clusters," and "Coverage ratio for absorbing adjacent small clusters." Raising the former two parameters reduces an insertion of a new breakpoint during clustering, and reducing the latter two parameters facilitates merging adjacent domains in the postprocessing phase, thereby in either case reducing the number of domain-splitting events. Alternatively, you can select the option "Without domain splitting" to skip this procedure.

6. You can create a user account before entering your data. To create a new account, select the "Create Account" tab and enter your email address and other information. Once registered, you can login to your account by selecting the "Login" tab. This is recommended for stable access to your data. By default (guest user mode), the MBGD server distinguishes different users using an HTTP cookie, which is dependent on the web browser you are using and will expire within 30 days.
7. The species code should not match the existing codes in MBGD. For this purpose, MBGD automatically adds "u_" before the input species code.
8. NCBI Taxonomy ID is used to locate your data in the taxonomy tree. Specify most closely related species if your species is not included in the NCBI Taxonomy database.
9. Here, you should check whether the protein sequence of each CDS (that is defined in the gene table) is correctly read and displayed. If there is no problem, the numbers of CDS and protein sequences displayed in the first table should agree with each other. You can check error messages by pressing the "View Error Message(s)" button that appears when the system detects errors.
10. Because My MBGD mode is the specialized mode for user-specific genome data, you may encounter a problem when using a precomputed ortholog table in My MBGD mode. To check whether you are currently in My MBGD mode, go to the top page and see whether the left-hand menu contains "Exit My MBGD mode." If this option exists, press it to exit. After exiting and reentering My MBGD mode, you can reload the clustering results once created in My MBGD mode by clicking the "Result Manager" tab on the My MBGD data input page (*see* Fig. 3a) and click the table ID in the cluster table list shown therein.
11. Even when you analyze the data for the second time or later, in which case you can simply open the clustering data to be analyzed without repeating **steps 1–3**, make sure that you have accessed MBGD at least once in your default browser and chosen the cluster table that you want to analyze beforehand. This step is required for accessing MBGD from the RECOG client program.
12. For filtering and coloring, multiple effects can be combined. For example, the effects of coloring include the base color (default: dark green), cells containing multiple genes, i.e., inparalogs (default: light green), genes hit by keyword search (default: pink), and neighboring clusters (assigned a different color for each cluster). The order of these effects and individual

color setting determine the final color assignment. Controlling the combination of various effects on the current ortholog table view using the control panel (Fig. 4, area 10) is essential for effective use of RECOG. Please see the manual available on the RECOG home page for details.

Acknowledgment

The author thanks Hirokazu Chiba, Hiroyo Nishide, and Motohiro Mihara for the development and maintenance of the MBGD Web service. The development of MBGD is supported by National Bioscience Database Center, Japan Science Technology Agency. Computational resources were provided by the Data Integration and Analysis Facility, National Institute for Basic Biology.

References

1. Fitch WM (1970) Distinguishing homologous from analogous proteins. Syst Zool 19:99–113
2. Pellegrini M, Marcotte EM, Thompson MJ, Eisenberg D, Yeates TO (1999) Assigning protein functions by comparative genome analysis: protein phylogenetic profiles. Proc Natl Acad Sci U S A 96:4285–4288
3. Marcotte EM, Pellegrini M, Ng HL, Rice DW, Yeates TO, Eisenberg D (1999) Detecting protein function and protein-protein interactions from genome sequences. Science 285:751–753
4. Overbeek R, Fonstein M, D'Souza M, Pusch GD, Maltsev N (1999) The use of gene clusters to infer functional coupling. Proc Natl Acad Sci U S A 96:2896–2901
5. Uchiyama I, Mihara M, Nishide H, Chiba H (2015) MBGD update 2015: microbial genome database for flexible ortholog analysis utilizing a diverse set of genomic data. Nucleic Acids Res 43:D270–D276
6. Uchiyama I (2003) MBGD: microbial genome database for comparative analysis. Nucleic Acids Res 31:58–62
7. Uchiyama I (2006) Hierarchical clustering algorithm for comprehensive orthologous-domain classification in multiple genomes. Nucleic Acids Res 34:647–658
8. Sonnhammer EL, Koonin EV (2002) Orthology, paralogy and proposed classification for paralog subtypes. Trends Genet 18:619–620
9. Chiba H, Uchiyama I (2014) Improvement of domain-level ortholog clustering by optimizing domain-specific sum-of-pairs score. BMC Bioinformatics 15:148
10. Uchiyama I (2007) MBGD: a platform for microbial comparative genomics based on the automated construction of orthologous groups. Nucleic Acids Res 35:D343–D346
11. Uchiyama I, Higuchi T, Kobayashi I (2006) CGAT: a comparative genome analysis tool for visualizing alignments in the analysis of complex evolutionary changes between closely related genomes. BMC Bioinformatics 7:472
12. Uchiyama I (2008) Multiple genome alignment for identifying the core structure among moderately related microbial genomes. BMC Genomics 9:515
13. Galperin MY, Makarova KS, Wolf YI, Koonin EV (2015) Expanded microbial genome coverage and improved protein family annotation in the COG database. Nucleic Acids Res 43: D261–D269
14. Kanehisa M, Goto S, Sato Y, Kawashima M, Furumichi M, Tanabe M (2014) Data, information, knowledge and principle: back to metabolism in KEGG. Nucleic Acids Res 42: D199–D205
15. Haft DH, Selengut JD, Richter RA, Harkins D, Basu MK, Beck E (2013) TIGRFAMs and genome properties in 2013. Nucleic Acids Res 41:D387–D395
16. Ashburner M, Ball CA, Blake JA, Botstein D, Butler H, Cherry JM, Davis AP, Dolinski K, Dwight SS, Eppig JT et al (2000) Gene ontology: tool for the unification of biology. The Gene Ontology Consortium. Nat Genet 25:25–29
17. Sievers F, Wilm A, Dineen D, Gibson TJ, Karplus K, Li W, Lopez R, McWilliam H,

Remmert M, Soding J et al (2011) Fast, scalable generation of high-quality protein multiple sequence alignments using Clustal Omega. Mol Syst Biol 7:539

18. Katoh K, Standley DM (2013) MAFFT multiple sequence alignment software version 7: improvements in performance and usability. Mol Biol Evol 30:772–780
19. Price MN, Dehal PS, Arkin AP (2010) FastTree 2—approximately maximum-likelihood trees for large alignments. PLoS One 5:e9490
20. Altschul SF, Madden TL, Schaffer AA, Zhang J, Zhang Z, Miller W, Lipman DJ (1997) Gapped BLAST and PSI-BLAST: a new generation of protein database search programs. Nucleic Acids Res 25:3389–3402
21. Smith TF, Waterman MS (1981) Identification of common molecular subsequences. J Mol Biol 147:195–197
22. Wu J, Kasif S, DeLisi C (2003) Identification of functional links between genes using phylogenetic profiles. Bioinformatics 19:1524–1530
23. Galperin MY, Koonin EV (2000) Who's your neighbor? New computational approaches for functional genomics. Nat Biotechnol 18:609–613
24. Drissi F, Merhej V, Angelakis E, El Kaoutari A, Carriere F, Henrissat B, Raoult D (2014) Comparative genomics analysis of Lactobacillus species associated with weight gain or weight protection. Nutr Diabetes 4:e109

Chapter 13

Exploring Protein Function Using the *Saccharomyces* Genome Database

Edith D. Wong

Abstract

Elucidating the function of individual proteins will help to create a comprehensive picture of cell biology, as well as shed light on human disease mechanisms, possible treatments, and cures. Due to its compact genome, and extensive history of experimentation and annotation, the budding yeast *Saccharomyces cerevisiae* is an ideal model organism in which to determine protein function. This information can then be leveraged to infer functions of human homologs. Despite the large amount of research and biological data about *S. cerevisiae*, many proteins' functions remain unknown. Here, we explore ways to use the *Saccharomyces* Genome Database (SGD; http://www.yeastgenome.org) to predict the function of proteins and gain insight into their roles in various cellular processes.

Keywords *Saccharomyces cerevisiae*, GO Term Finder, Protein function, Gene Ontology, YeastMine

1 Introduction

The *Saccharomyces* Genome Database (SGD; http://www.yeastgenome.org) is the premier website for researchers interested in curated data about the budding yeast, *Saccharomyces cerevisiae* [1]. SGD provides comprehensive, curated biological information regarding protein function, cellular location, interactions, and expression, as well as information on mutant phenotypes for *S. cerevisiae* along with search and analysis tools to explore these data, thereby enabling the discovery of functional relationships between sequence and gene products in fungi and higher eukaryotic organisms. The tools provided by SGD leverage this expertly curated information. For predicting protein function, users can access the wealth of curated information available on SGD's locus summary pages, search and retrieve information using YeastMine, or analyze gene function using Gene Ontology (GO) enrichment tools.

For researchers interested in a single, specific protein, the Locus pages at SGD contain function summaries, gene ontology

Daisuke Kihara (ed.), *Protein Function Prediction: Methods and Protocols*, Methods in Molecular Biology, vol. 1611,
DOI 10.1007/978-1-4939-7015-5_13, © Springer Science+Business Media LLC 2017

annotations, protein-protein interaction information, as well as other curated, published experimental data to help predict a protein's role in cellular processes. Additionally, more specific protein information, such as predicted domains, can be found on Protein pages. This information can shed light on a protein's function, and help direct experimental research aimed at determining a protein's function.

By leveraging these curated data, users can also predict the functions of similar proteins using the GO Term Finder enrichment tool at SGD. The GO project (http://geneontology.org/) is a collaborative initiative involving many model organism databases, and was initiated by SGD, FlyBase (*Drosophila*), and Mouse Genome Informatics (MGI) to provide consistent descriptions of gene products across various organisms and databases [2]. GO has developed and maintains three structured ontologies to describe the biological processes, molecular functions, and cellular locations of gene products. GO Term Finder performs enrichment analysis by taking lists of genes as input, then looking for overrepresented ontology terms based on the associated annotations compared to all other annotations [3]. This enables the identification of biological processes, molecular functions, and cellular locations that the gene products in the input list may have in common.

Finally, YeastMine is a powerful search tool that can be queried to identify lists of proteins based on similarity in function, or other shared characteristics, to gain insight into biological processes [4]. YeastMine contains all curated information stored in the SGD database and more. Pregenerated queries can be employed by users interested in a specific process to find novel players, or to identify common functions within a group of proteins to predict function or involvement in a specific cellular process.

Individually or together, the various tools at SGD can provide insight into the cellular processes in which proteins are involved, thereby supporting the development of a fully defined model of the eukaryotic cell.

2 Materials

Using any of the tools mentioned above requires an internet-connected computer with a web-browser to connect to the *Saccharomyces* Genome Database (http://www.yeastgenome.org).

3 Methods

The data in SGD are updated on a daily basis as more information is curated. Therefore, results for these methods and case studies outlined below may be slightly different than described.

3.1 Summary Pages

1. Open the URL, www.yeastgenome.org, in any modern web browser (e.g., Chrome, Firefox, or Safari).
2. Enter a gene name (e.g., NOG2) or systematic name (e.g., YNR053C) into the search box on the top right-hand side of the page and hit return.
3. The resulting NOG2/YNR053C Summary page has many sections that can be used to gain insight into the protein's potential function. The top Overview section contains a Description that summarizes curated information. The Gene Ontology (GO) section contains GO terms used to annotate the gene product, and also a brief summary of its function, cellular location, and biological processes in which it is involved (Fig.1a). GO terms are linked to a term-specific page that displays the term's definition as well as a list of other co-annotated gene products. Proteins with similar functions or that are involved in the same processes have closely related annotations, thus GO annotations of characterized proteins can be leveraged to predict the functions of uncharacterized proteins.
4. Based on the premise that proteins acting in the same pathway or involved in the same process may share mutant phenotypes, the data in the Phenotype section can also provide insight into a protein's function. The Phenotype section contains a summary of the curated experimental phenotype data, as well as the individual mutant phenotype annotations (Fig. 1b).
5. Protein function can also be predicted based on the proteins that interact with it (*see* **Note 1**). The Interaction section of this page summarizes the methods and number of physical and genetic interactions that have been identified for this locus (Fig. 1c). Selecting the "Details" link on the right-hand side of any section will provide access to more detailed information, such as a list of the interacting genes and proteins for the "Interaction Details" link. By accessing a list of interacting proteins, available via the "Analyze" link on the details page, the GO Enrichment tool described below can be used to predict protein function.
6. For additional insight into a protein's function, click on "Protein Details" on the top right link of the "Protein" section of the page, or the "Protein" tab at the top of the page. The "Protein" page contains the following types of protein-specific information: cellular abundance, predicted domains, protein sequence, posttranslational modifications, and physical properties. The predicted domains can help predict a protein's function. For example, Nog2p, a putative GTPase, has predicted GTP-binding protein domains in multiple locations (Fig. 2).

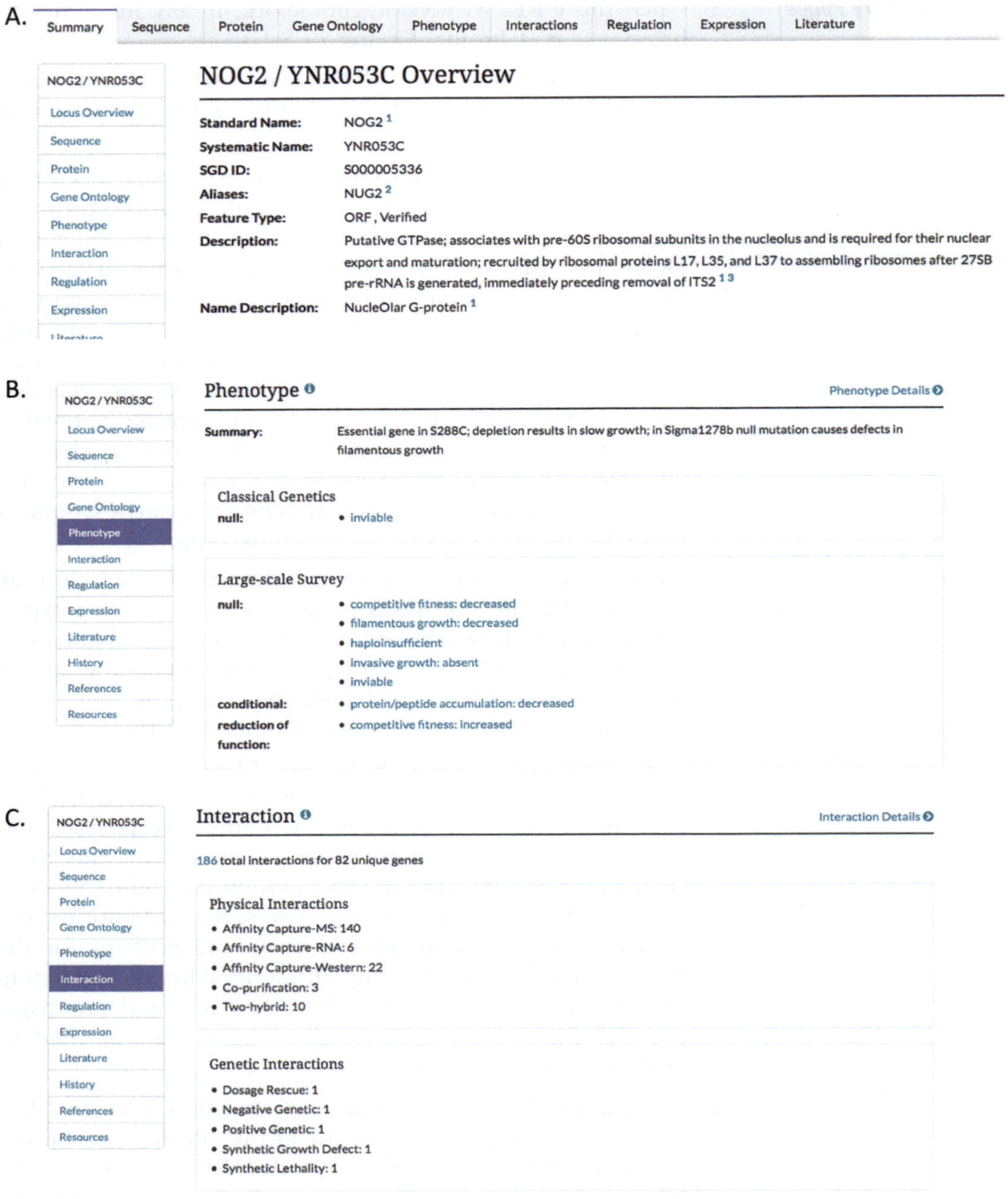

Fig. 1 Summary of information on a locus summary page. (**a**) NOG2/YNR053C Overview section has a description of known protein function information. (**b**) Phenotype information can help predict a protein's function. (**c**) Interaction data can be used to predict protein function and process involvement

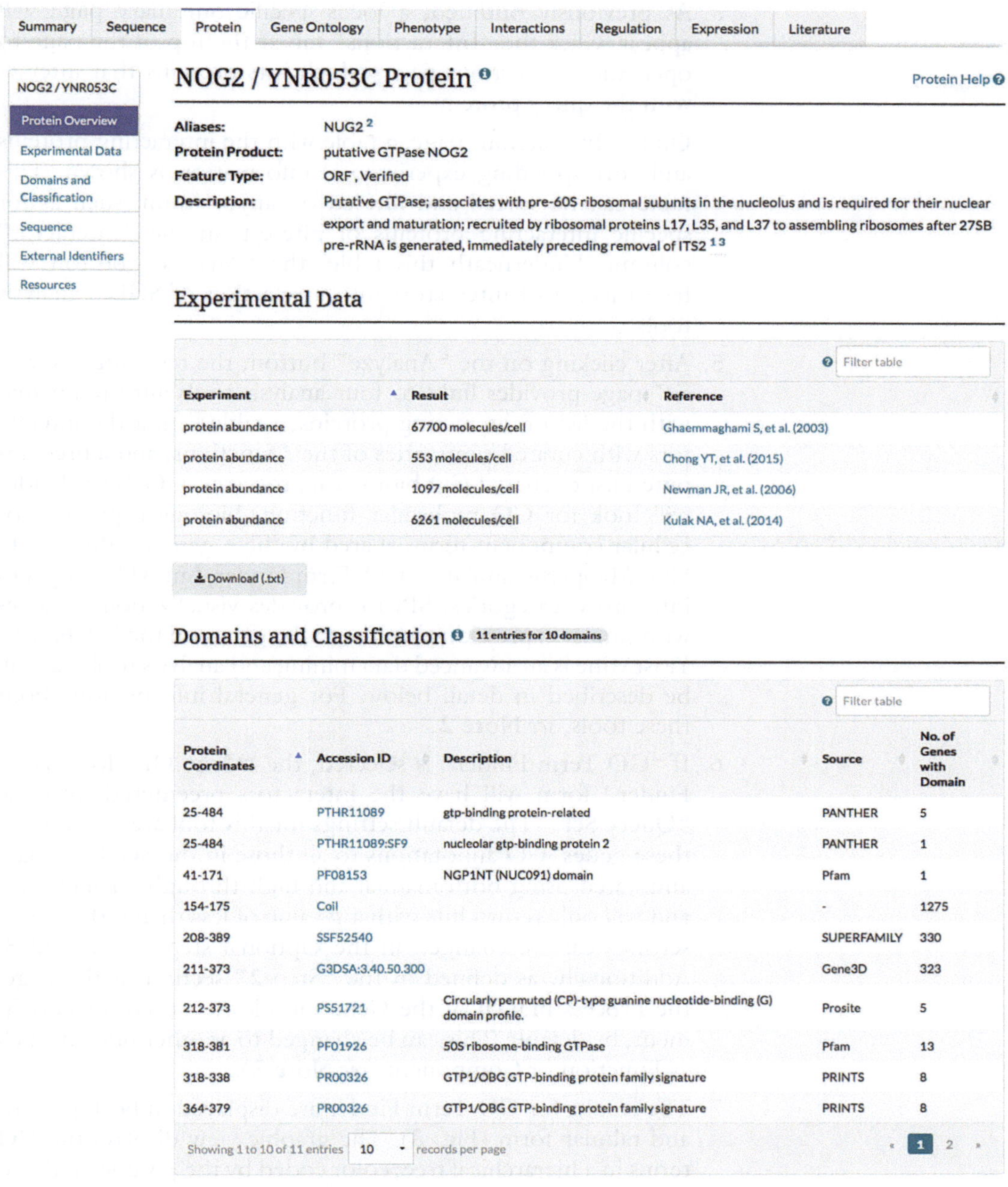

Fig. 2 NOG2 protein page. Protein abundance and domains on the protein page can be used to predict protein function

3.2 Gene Ontology Enrichment Tool

1. Open the URL www.yeastgenome.org in any modern web browser (e.g., Chrome, Firefox, or Safari).
2. Enter a gene name in the search box as described above, such as YAL044W-A.

3. As previously outlined, a locus-specific Summary page will appear. Click the "Interactions" tab at the top of the page to open the Interactions page, which lists proteins that interact with the query protein.
4. On the Interactions page, a table with the interacting proteins and corresponding experimental information is shown. This table can be sorted and filtered on any column, such as for specific interacting proteins of interest, in the "Interactor" column. Underneath this table, the "Analyze" button will feed this list of interactors into a selection of SGD's analysis tools.
5. After clicking on the "Analyze" button, the resulting "Analysis" page provides links to four analysis tools auto-populated with the list of interacting proteins, a table listing the interactors with curated summaries of their functions, and a precomputed list of shared GO biological processes. GO Term Finder will look for GO molecular function, biological process, or cellular component terms shared by these gene products. GO Slim Mapper is similar to GO Term Finder, but will sort genes into broad categories. SPELL provides visualization of genes with similar expression patterns to the genes on the list. Finally, YeastMine is an advanced data mining and analysis tool that will be described in detail below. For general information about these tools, *see* **Note 2**.
6. If "GO Term Finder" is selected, the "Gene Ontology Term Finder" form will have the interactors preentered into the "Query Set." The default settings for this tool are to compare these genes' GO annotations to all those in the SGD database annotated using both manual and high-throughput methods, and will only return hits with a p-value of less than 0.01. These settings can be changed in the Optional steps: 3, 4, and 5. Additionally, as defined in the "Step 2" section of the page, the Process branch of the Gene Ontology is used for enrichment, by default. This can be changed to another branch, such as Function or Component (*see* **Note 3**).
7. The results for GO Term Finder are displayed in both graphic and tabular form (Fig. 3). The graphic view displays the GO terms in a hierarchical tree; color coded by the p-value, and also indicates the genes that are annotated to the specific term. The table view displays the GO terms to which the genes are directly annotated, as well as parent terms, along with cluster and background frequencies and p-values. The results can also be downloaded using the "Download Full Results" link at the bottom of the results table. Links to GO term pages, as well as to the GO consortium's search and browse tool (AmiGO; http://amigo.geneontology.org/amigo), are provided within the table of results.

A.

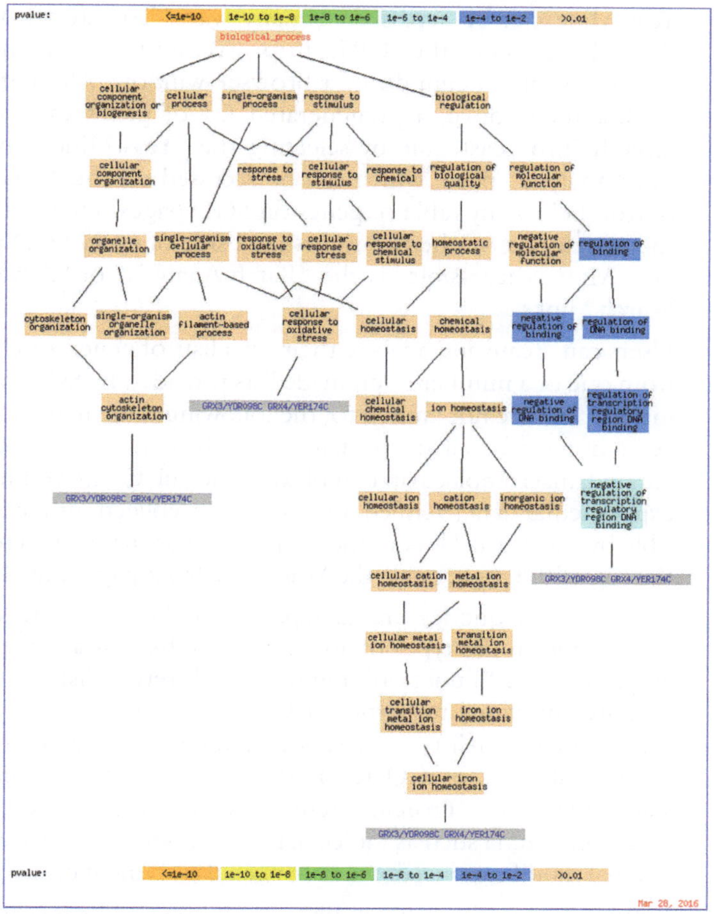

B.

Terms from the Process Ontology

Gene Ontology term	Cluster frequency	Background frequency	P-value	FDR	Expected False Positives	Genes annotated to the term
negative regulation of transcription regulatory region DNA binding | AmiGO	2 out of 7 genes, 28.6%	2 out of 7165 background genes, 0.0%	8.42e-05	0.00%	0.00	GRX3/YDR098C, GRX4/YER174C
negative regulation of DNA binding | AmiGO	2 out of 7 genes, 28.6%	3 out of 7165 background genes, 0.0%	0.00025	0.00%	0.00	GRX3/YDR098C, GRX4/YER174C
negative regulation of binding | AmiGO	2 out of 7 genes, 28.6%	4 out of 7165 background genes, 0.1%	0.00051	0.00%	0.00	GRX3/YDR098C, GRX4/YER174C
regulation of transcription regulatory region DNA binding | AmiGO	2 out of 7 genes, 28.6%	5 out of 7165 background genes, 0.1%	0.00084	0.00%	0.00	GRX3/YDR098C, GRX4/YER174C
regulation of DNA binding | AmiGO	2 out of 7 genes, 28.6%	7 out of 7165 background genes, 0.1%	0.00177	0.00%	0.00	GRX3/YDR098C, GRX4/YER174C
regulation of binding | AmiGO	2 out of 7 genes, 28.6%	12 out of 7165 background genes, 0.2%	0.00554	0.00%	0.02	GRX3/YDR098C, GRX4/YER174C

Download Full Results

Fig. 3 GO Term Finder results. (**a**) Graphic view of GO term enrichment of proteins that interact with Yal044w-ap. (**b**) Table view of GO Term Finder results for Yal044w-ap interacting proteins

3.3 YeastMine

1. YeastMine can be accessed in many ways. To directly access YeastMine, open the URL http://yeastmine.yeastgenome.org/yeastmine/begin.do in a browser window. Alternatively, as described above, a pregenerated list of genes can be fed directly into YeastMine by selecting the "YeastMine" link on the "Analysis" page, which can be accessed via the "Analyze" button below any table of genes on SGD pages. Finally, below the search box on the top-right-hand side of every SGD page, YeastMine is accessible via the "Batch Analysis" or "Advanced Search" links.
2. Users can create and analyze their own lists of genes, or choose from one of a number of premade lists provided by SGD, which include, but are not limited to, the following: human genes with yeast homologs, all uncharacterized ORFs in *S. cerevisiae*, all curated macromolecular complexes, and all ORFs verified by experimental data (Verified_ORFs). These collections are accessible by the "Lists" link in the purple toolbar that runs across at the top of the page, or on the YeastMine front page (Fig. 4b).
3. Pregenerated queries are grouped into different categories, depending on the types of annotated results that are analyzed (Fig. 4c). The "Function" templates will return lists of genes that are annotated to particular GO terms, or pathways. Conversely, for a given list of genes, other Function queries will return all the pathways or GO terms that are annotated within the starting list. The "Proteins" templates will return lists of genes based on criteria such as molecular weight, sequence, posttranslational modifications, domains, or cellular abundance.
4. Returned lists of results can be further screened by looking for similarities within other curated data, such as interactions and phenotypes, as outlined below in Subheading 4.
5. In addition, lists can be saved by creating a "MyMine" account, using the "Log In" link found on the top right-hand side of the header (Fig. 4a).

4 Case Studies

4.1 Predicting the Function of an Uncharacterized Protein with GO Term Finder

In this example, GO Term Finder will be used to predict the putative function of YAL044W-A, a protein of unknown function. As outlined in Subheading 3, after entering YAL044W-A in the search box, a page summarizing all curated information about YAL044W-A will be returned. YAL044W-A is noted to be similar to a *Schizosaccharomyces pombe* putative DNA repair protein, but no other functional information is known. However, it has been shown to physically interact with seven unique proteins (Fig. 5).

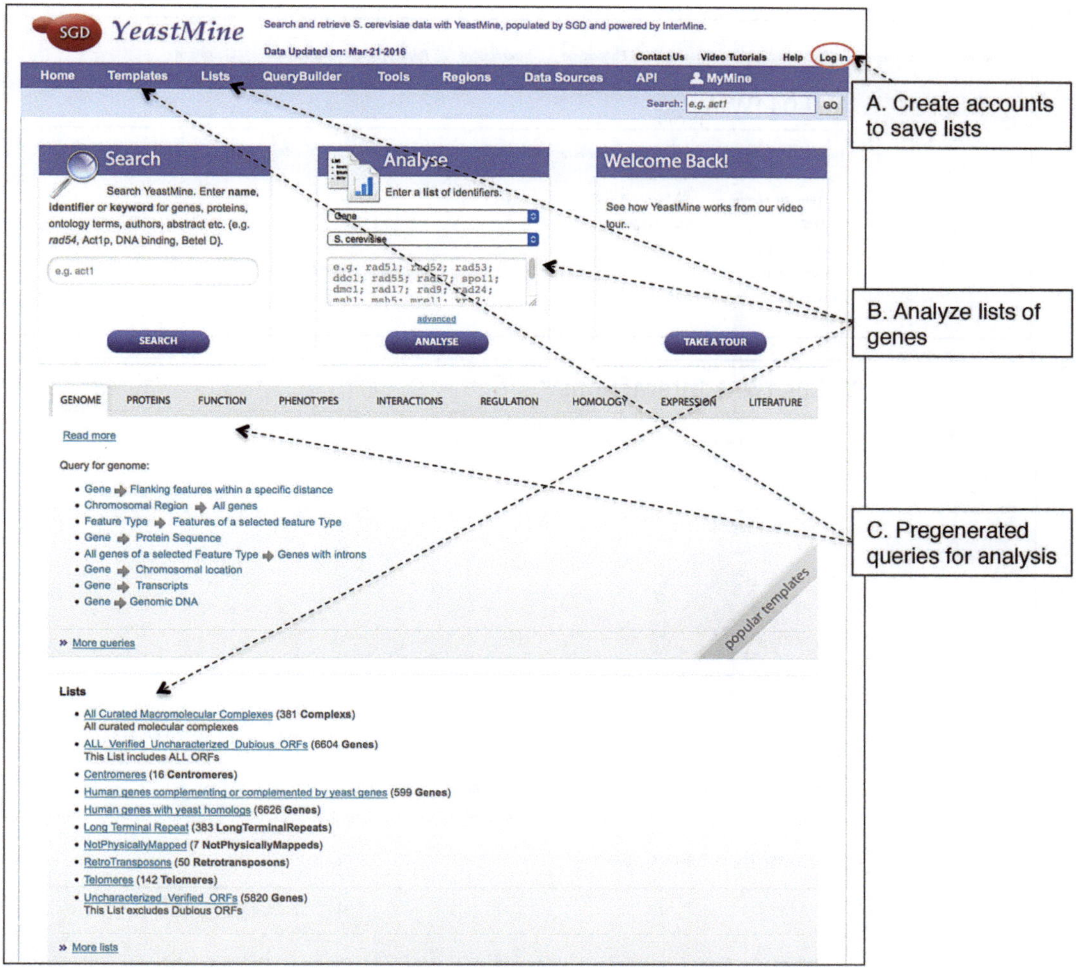

Fig. 4 Analysis tools that are available on the YeastMine home page

Researchers can gain insight into the putative biological function of the uncharacterized gene, YAL044W-A, by leveraging the curated information about these interacting proteins' functions with GO Term Finder to determine GO terms that have been annotated to this input set.

First, to identify known proteins that interact with YAL044W-A, click either on the "Interactions" tab at the top of the page, or the "Interactions Details" link in the "Interactions" section of the page. In the "Annotations" section, a list of seven proteins that interact with Yal044w-ap, along with experimental details and references, are listed in a table (Fig. 5b). An "Interaction Network" below the "Annotations" section is a graphic display of the interactions of these proteins. In this example, Grx3p/Ydr098cp, Grx4p/Yer174cp, Mgs1p/Ynl218wp, Mpa43p/Ynl249cp,

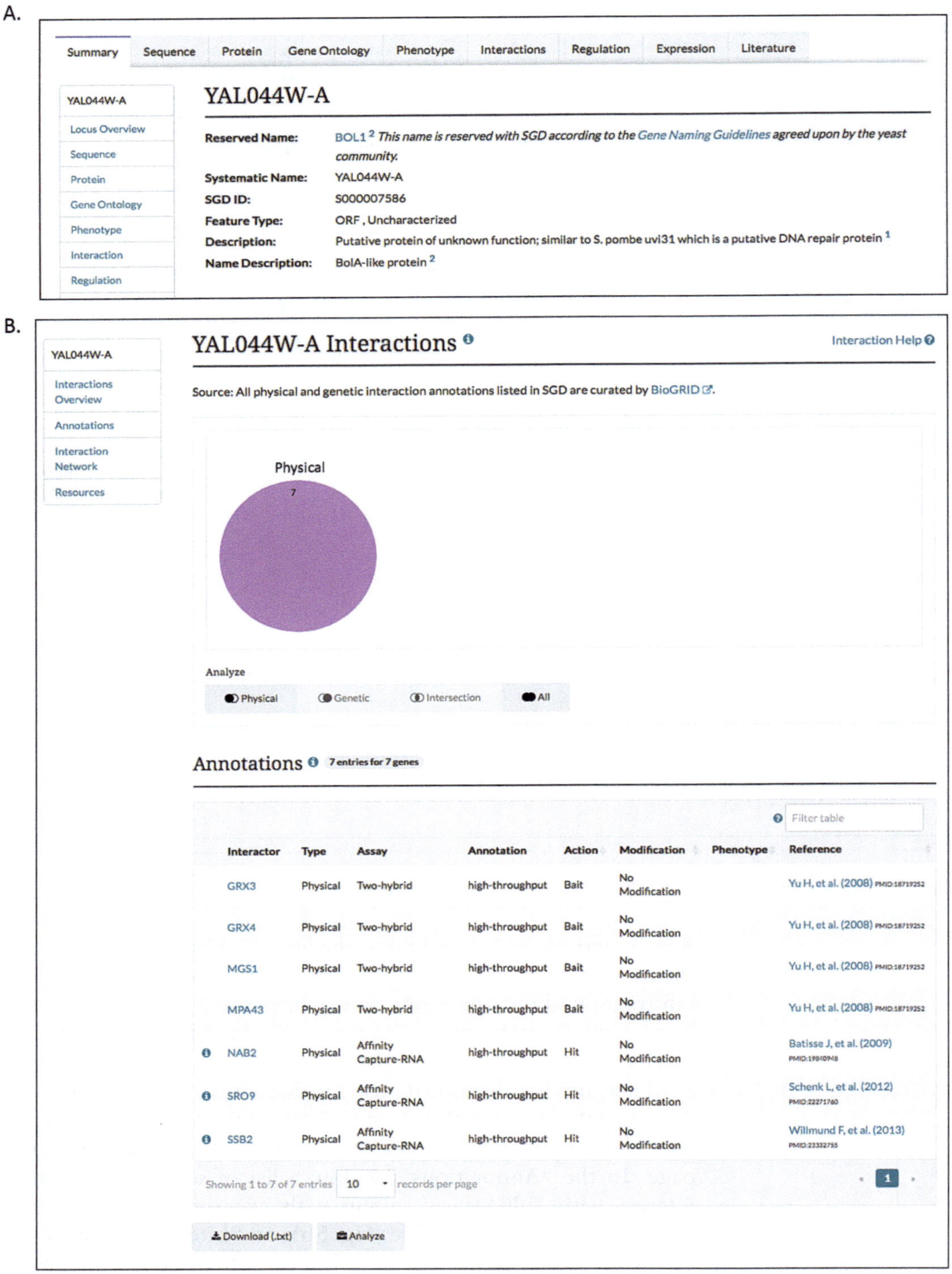

Fig. 5 Curated information about YAL044W-A. (**a**) Description of YAL044W-A, an uncharacterized ORF. (**b**) List of proteins that physically interact with Yal044w-ap

Nab2p/Ygl122cp, Sro9p/Ycl037cp, and Ssb2p/Ynl209wp all have been shown to physically interact with Yal044w-ap.

Once identified, the annotations for these interactors can be used with the GO enrichment tool to determine YAL044W-A's putative function. Directly below the annotations table, select the "Analyze" button to pass these gene names to SGD's analysis tools. On this "Analyze" page, GO Term Finder, GO Slim Mapper, SPELL, and YeastMine are listed above the input list of genes. This list of interactors also includes the curated description of their functions, which could also shed some light on the function of the gene of interest. To find common GO terms to which these genes have been annotated, this list can be sent directly to GO Term Finder via the "GO Term Finder" link above the list of interactors.

In this case, the GO Term Finder form will be prefilled with GRX3/YDR098C, GRX4/YER174C, MGS1/YNL218W, MPA43/YNL249C, NAB2/YGL122C, SRO9/YCL037C, and SSB2/YNL209W. To predict the function of YAL044W-A, the researcher would enrich for "Process" annotations, which describe broad biological concepts (the default selection). In this case study, all default settings would be used. After clicking "Search," the GO Term Finder results are displayed both graphically and in tabular form (Fig. 3). Within the graphical display of the results, the hierarchical structure of GO can be observed (Fig. 3a). Based on the enrichment of GO terms in the input list of interactors, YAL044W-A is predicted to have a role in regulating transcription with GRX3 and GRX4, likely through negative regulation of DNA binding (http://yeastgenome.org/go/GO:2000678/overview). Additionally, the table displays the calculated frequency of GO terms in the cluster (input list) and the background list of genes, the entire yeast genome in this case. The *p*-value of 8.42e-05 and expected false positive of 0.00 indicate a high confidence in this GO term process enrichment (Fig. 3b). The enrichment of processes, such as this case study, can provide direction toward experimentally determining a protein's function.

4.2 Identifying Novel Components of the Mitochondrial Ribosome

In this example, YeastMine will be used to identify novel, potential subunits of the mitochondrial ribosome. In yeast, there are 91 genes annotated to the mitochondrial ribosome or a ribosomal subunit, less than half of what has been found for cytosolic ribosomes. Leveraging the wealth of curated information in SGD, as well as the power of YeastMine's search and analysis, a list of putative novel mitochondrial ribosomal subunits can be identified for future lab experiments. First, the researcher could identify all known mitochondrial ribosome proteins. To identify novel proteins, this list can be used to generate a second list of proteins that interact with the known mitochondrial ribosome proteins. Finally, restricting the list of interactors to uncharacterized proteins would reduce the list to a manageable target list of possible genes of interest.

To identify all the genes encoding subunits of the mitochondrial ribosome, researchers would use the "Function" template query called "GO Term name [and children of this term] → All genes" (Fig. 6b). To keep the search broad, the constraints "CONTAINS" "mito* ribosom*" would be used. This returns a list of 91 unique genes, annotated to GO terms such as "mitochondrial ribosome," "mitochondrial large ribosomal subunit," and "mitochondrial small ribosomal subunit." This list should be saved and named using the "Save as List" button above the results table.

The second step would be used to identify all proteins that interact with this list of ribosomal subunits. To do this, the user would use the "Gene → Interaction" query found under the "Interaction" category in Templates. Rather than looking up the interactions for a single gene, the checkbox next to "constrain to be IN saved Gene list" would be selected, and then the list created and saved in the previous step would be selected from the dropdown. This would return 3478 interactions involving 1030 unique genes (listed as Participant 2 Standard Name) for the original list. Again, this list of interacting proteins would be named and saved using the "Save as List" button above the results table.

This list of proteins that interact with known mitochondrial ribosome proteins can be filtered against a pregenerated list of uncharacterized genes. To do this, the researcher would select "Lists," found at the top of the page, next to "Templates." Under lists, the "View" tab will display all the SGD pregenerated lists as well as any personal lists that the researcher created. First, the recently created list of proteins that interact with known mitochondrial ribosomal subunits and the "Uncharacterized_ORFs" list of 672 genes are selected. Then, selecting "Intersect" under "Actions:" above all the lists will return a list of 34 genes that are uncharacterized and have been shown to interact with known mitochondrial ribosomal subunits. This list of 34 genes would be a good starting point for experimentally confirming which of these predicted mitochondrial ribosomal subunits were actual mitochondrial ribosomal subunits. A detailed video description of this case study can be found on the SGD YouTube channel, here: https://www.youtube.com/watch?v=AoVfmaZmTK4. To learn more about the many features of YeastMine, visit the SGD YouTube channel (https://www.youtube.com/SaccharomycesGenomeDatabase) and enter "YeastMine" in the search box at the top of the page.

5 Notes

SGD provides a number of tips and help for the resources mentioned above.

1. General help about curated protein function information at SGD can be found on the function help page: http://www.yeastgenome.org/help/function-help.

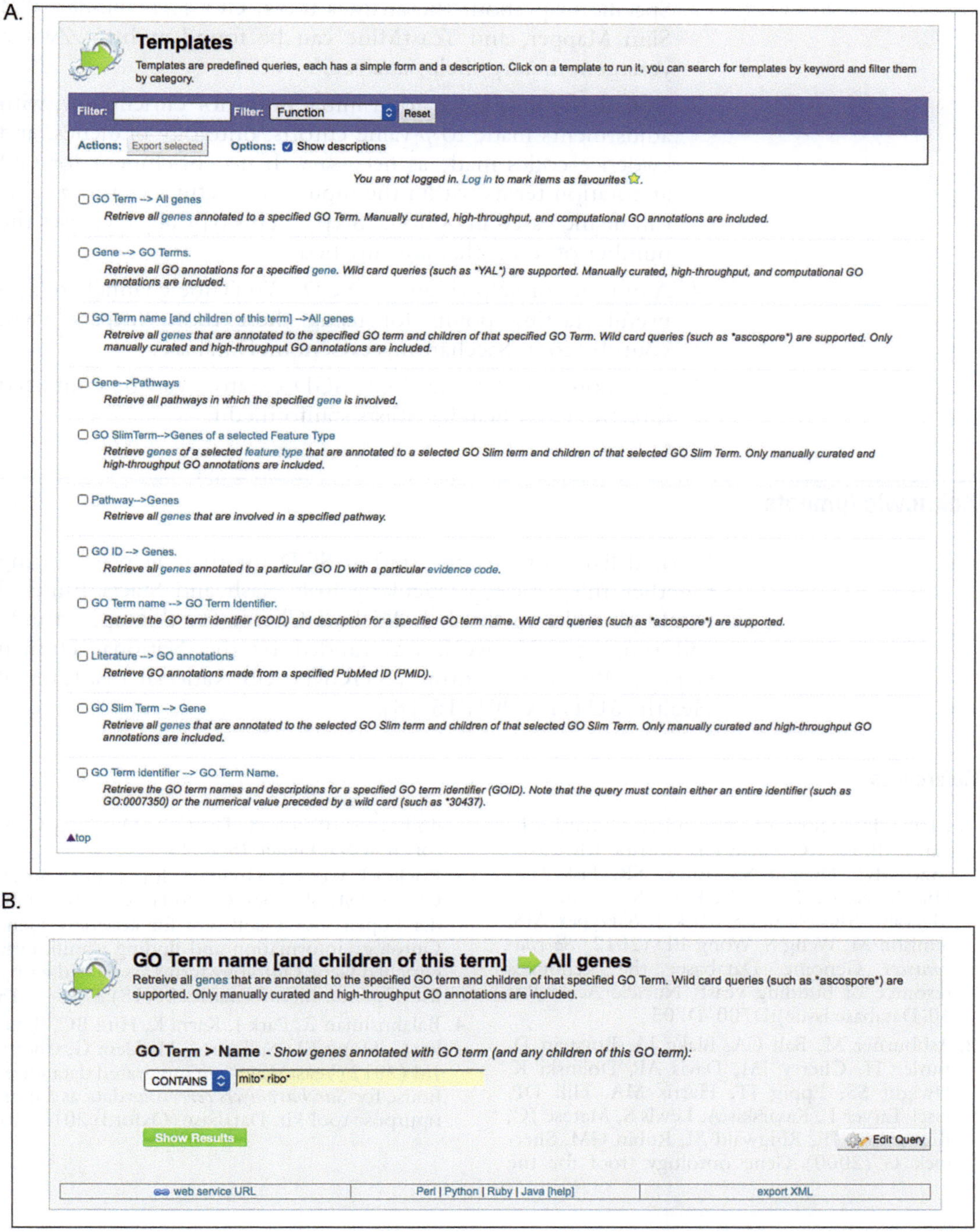

Fig. 6 Function queries in YeastMine for predicting protein function. (**a**) A list of pregenerated queries available for examining protein function. (**b**) Using specific query to identify all known mitochondrial ribosome proteins

2. Specific help about the analysis tools, GO Term Finder, GO Slim Mapper, and YeastMine can be found at http://www.yeastgenome.org/help/analyze/.
3. Default settings are a good initial point for enrichment, with adjustments made to *p*-value cutoffs, ontology branches, and evidence codes made as necessary. If no enrichment for GO annotation terms within the input list is returned, the *p*-value can be increased in optional Step 5. This typically increases the number of terms that are enriched.
4. A number of help videos on SGD's YouTube channel are also useful starting points for using these tools: https://www.youtube.com/SaccharomycesGenomeDatabase.
5. For more specific questions, SGD curators can be contacted directly at sgd-helpdesk@lists.stanford.edu.

Acknowledgments

I would like to thank the staff at SGD for their help in putting together this article, particularly Rob Nash and Stacia Engel. I would also like to thank J. Michael Cherry for this opportunity and support. This work was funded by the National Human Genome Research Institute (NHGRI), US National Institutes of Health [5U41HG001315-18].

References

1. Cherry JM, Hong EL, Amundsen C, Balakrishnan R, Binkley G, Chan ET, Christie KR, Costanzo MC, Dwight SS, Engel SR, Fisk DG, Hirschman JE, Hitz BC, Karra K, Krieger CJ, Miyasato SR, Nash RS, Park J, Skrzypek MS, Simison M, Weng S, Wong ED (2012) *Saccharomyces* Genome Database: the genomics resource of budding yeast. Nucleic Acids Res 40(Database issue):D700–D705
2. Ashburner M, Ball CA, Blake JA, Botstein D, Butler H, Cherry JM, Davis AP, Dolinski K, Dwight SS, Eppig JT, Harris MA, Hill DP, Issel-Tarver L, Kasarskis A, Lewis S, Matese JC, Richardson JE, Ringwald M, Rubin GM, Sherlock G (2000) Gene ontology: tool for the unification of biology. The Gene Ontology Consortium. Nat Genet 25(1):25–29
3. Boyle EI, Weng S, Gollub J, Jin H, Botstein D, Cherry JM, Sherlock G (2004) GO::TermFinder—open source software for accessing Gene Ontology information and finding significantly enriched Gene Ontology terms associated with a list of genes. Bioinformatics 20(18):3710–3715
4. Balakrishnan R, Park J, Karra K, Hitz BC, Binkley G, Hong EL, Sullivan J, Micklem G, Cherry JM (2012) YeastMine—an integrated data warehouse for *Saccharomyces cerevisiae* data as a multipurpose tool-kit. Database (Oxford) 2012:201

Chapter 14

Network-Based Gene Function Prediction in Mouse and Other Model Vertebrates Using MouseNet Server

Eiru Kim and Insuk Lee

Abstract

The mouse, *Mus musculus*, is a popular model organism for the study of human genes involved in development, immunology, and disease phenotypes. Despite recent revolutions in gene-knockout technologies in mouse, identification of candidate genes for functions of interest can further accelerate the discovery of novel gene functions. The collaborative nature of genetic functions allows for the inference of gene functions based on the principle of guilt-by-association. Genome-scale co-functional networks could therefore provide functional predictions for genes via network analysis. We recently constructed such a network for mouse (MouseNet), which interconnects over 88% of protein-coding genes with 788,080 functional relationships. The companion web server (www.inetbio.org/mousenet) enables researchers with no bioinformatics expertise to generate predictions that facilitate discovery of novel gene functions. In this chapter, we present the theoretical framework for MouseNet, as well as step-by-step instructions and technical tips for functional prediction of genes and pathways in mouse and other model vertebrates.

Keywords Functional prediction, Mouse, Model vertebrates, Co-functional network, Guilt-by-association, Network biology

1 Introduction

Understanding human physiology requires functional characterization of biomolecules. Proteins are the major functional molecules for the execution of biochemical processes and support of cellular and organismal structures. The consensus coding sequence database (CCDS) [1] lists approximately 20,000 human protein-coding genes, and 40% of these are not yet annotated by Gene Ontology biological process (GOBP) terms with experimental evidence [2]. In addition, most human genes are pleiotropic. Laboratory organisms have been widely used to study human biology due to the technical and ethical limitations of experimental analysis in humans. *Mus musculus*, the laboratory mouse, is a popular model organism in biomedical research because of its similarities to humans in terms of genome content and physiology. More than 17,000 mouse

Daisuke Kihara (ed.), *Protein Function Prediction: Methods and Protocols*, Methods in Molecular Biology, vol. 1611, DOI 10.1007/978-1-4939-7015-5_14, © Springer Science+Business Media LLC 2017

protein-coding genes have human orthologs (~85% of the human-coding genome) based on the annotation (as of March 2016) of the Mouse Genome Informatics (MGI) database [3]. Because functional defects in human orthologs frequently result in disease-associated phenotypes in mouse, functional data on mouse proteins can be valuable for disease research.

Loss-of-function (LoF) analysis is commonly used to study protein function in laboratory animals. Due to the many-to-many relationship between genes and functions, exhaustive searches among all possible gene–function relationships are impractical. It is therefore necessary to reduce the search space by selecting more probable gene–function associations using high-throughput screening or bioinformatics analysis. Various types of information, such as sequence homology, expression, molecular interactions, and synteny, have been used to predict gene–function associations. Recently, co-functional networks have become popular in gene function prediction in a wide variety of organisms. Most, if not all, proteins work together to perform biological processes. With a map of functional relationships between genes, we can infer gene functions from those of related genes. This guilt-by-association principle has proven highly effective in the study of human biology using not only humans [4] but also laboratory animals such as worm [5] and fly [6].

Functional associations between coding genes can be inferred from molecular interactions. For example, a physical interaction between proteins suggests functional coupling. There are dozens of experimental methods for detecting protein–protein interactions (PPI) in vivo and in vitro [7]. Although this molecular interaction approach has made significant contributions to the mapping of PPIs, not all functional associations are based on physical interactions between proteins. For instance, similarities in genomic context [8, 9] or transcriptional regulation [10] can also reflect functional associations between genes. In addition, functional links in one species can be inferred from those between orthologs in other species [11]. Functional links based on these different methods tend to be complementary, and integration using a Bayesian statistics framework generally results in substantial increases in the accuracy of links as well as coverage of the coding genome [12], both of which significantly influence the success of network-based functional prediction.

Co-functional networks of mouse genes have been developed by multiple research groups through the MouseFunc competition [13]. Recently, we improved one of the networks, MouseNet (www.functionalnet.org/mousenet) [14], by incorporating a large amount of public genomics data that were not allowed in the competition setting. For example, more than 76,000 mouse mRNA expression samples from Gene Expression Omnibus (GEO) [15] were subject to co-expression analysis, while fewer than 250 were used for network inference during the MouseFunc competition. For the creation of the expanded network, we inferred a total of 183 co-expression networks from 183 GEO series (GSE) comprising 8,154 microarray experiments. Integration of these 183 co-expression networks resulted in 180,037 functional couplings among 14,087 mouse genes. In addition to the gene expression data, other types of co-functional link data have accumulated since development of the previous version of network. The updated mouse co-functional network, MouseNet v2 [16], constructed by integrating 13 distinct data types, contains 788,080 functional couplings among 17,714 mouse genes (>88% of coding genome). Significant improvements to MouseNet v2's network-based predictions [16] derive from incorporation of associalogs—orthology-based transferred functional links [11]—from co-functional networks for yeast [17], fly [6], and human [18]; mouse PPIs compiled from a meta database [19]; and links inferred from improved genomic context methods [20, 21].

Here, we present the MouseNet v2 companion web server (http://www.inetbio.org/mousenet) and demonstrate how functional predictions for mouse genes and pathways are generated via two alternative network search options: pathway-centric or gene-centric network search. The network searches return various prediction results to assist researchers in making decisions regarding the design of follow-up functional analyses. We therefore describe how these search results can be interpreted. For researchers who are interested in analysis with particular data types, the web server provides all 13 component networks individually from the download page. Researchers may also need to use a co-expression network for a particular biological context corresponding to a specific GSE. Thus, co-expression networks for each of the 138 GSEs are available from the web server. Lastly, in addition to mouse, MouseNet v2 provides predictions for eight other vertebrates (rat, chimpanzee, *Rhesus macaque*, dog, cattle, chicken, western clawed frog, and zebrafish) that are useful laboratory models for human biology. High homology between mouse and these eight vertebrates allowed for this extension through orthology-based transfer of network information.

2 Materials

2.1 Input Data

There are two alternative network search options available for functional predictions: (1) pathway-centric network search, where new members of a pathway are found based on network direct neighbors and (2) gene-centric network search, where functions are inferred from network neighbors. Both options accept gene names—either Entrez gene identification numbers (Entrez ID) or gene symbols (*see* **Note 1**)—as input to start a network search, though the roles of the input genes are quite different in the two search paths.

In pathway-centric network search, input genes act as *guide genes* that provide seed information to be propagated through the network. Thus, genes that receive a significant amount of transferred functional information from the guide genes become candidates for the same function. In this case, reliability of seed information significantly affects prediction accuracy. Functional annotations for mouse genes are available from various pathway databases, such as GOBP [2] and the Kyoto Encyclopedia of Genes and Genomes (KEGG) [22]. GOBP is particularly useful for its comprehensiveness and evidence codes that can be leveraged in selecting guide genes (*see* **Note 2**). As diseases are closely associated with functions, disease annotation databases, such as Online Mendelian Inheritance in Man (OMIM) [23], Disease ontology (DO) [24], Genome-wide association study (GWAS) catalog [25], and GWASdb [26] (*see* **Note 3**), can also provide guide genes. Pathway-centric network searches can be initiated not only with mouse genes but also with genes from any of the eight other model vertebrates. If users submit gene names for organisms other than mouse, the corresponding organism name must be selected from the menu. Since the mouse gene network is still used to conduct the search, input genes from other vertebrates are automatically converted into mouse orthologs using MGI vertebrate homology information before the search begins.

In gene-centric network search, input genes act as *query genes* in the search. Thus, input for this search option could be any valid gene from mouse or any of the eight other vertebrates. For each submission, organism selection and gene name conversion proceed as in pathway-centric network search.

2.2 Website

The MouseNet v2 companion web server is located at http://www.inetbio.org/mousenet and includes multiple services that are available from different pages. The tutorial page includes step-by-step instructions and screenshots for running functional predictions using the web server. The network-search page links to input data submission pages for conducting functional predictions. The network-download page provides various network edge information for free.

3 Methods

As mentioned above, MouseNet provides two alternative network search paths on the network-search page: (1) pathway-centric search for predicting novel genes in a pathway and (2) gene-centric search for predicting a gene's functions and associated phenotypes. Principles and step-by-step procedures for each prediction path are described below.

3.1 Predicting Novel Genes for a Pathway or Phenotype

Our pathway view is still largely limited and remains to be improved through identification of new member genes in each pathway. The use of the guilt-by-association principle for functional networks is a plausible approach for identifying novel genes in a pathway because pathway member genes are highly likely to be functionally coupled. In other words, if some pathway genes are known, their neighbors in a functional network may also be involved in the same pathway. In addition, the majority of phenotypes, such as human diseases, are regulated at the pathway level. Therefore, we may employ the guilt-by-association principle to predict candidate genes for phenotypes. In a genome-scale functional network, a set of pathway genes may have many network neighbors that can be prioritized by the total strength of connections to the pathway genes measured by edge weight. For example, functional links in MouseNet are given log likelihood scores (LLS) [12] (*see* **Note 4**), and neighbors of a set of pathway genes (i.e., guide genes) are ranked by the sum of the LLS of edges to all pathway genes. The whole procedure and results for pathway-centric searches are summarized in Fig. 1.

On the submission page for pathway-centric network searches, one may initiate a search by submitting guide genes, which

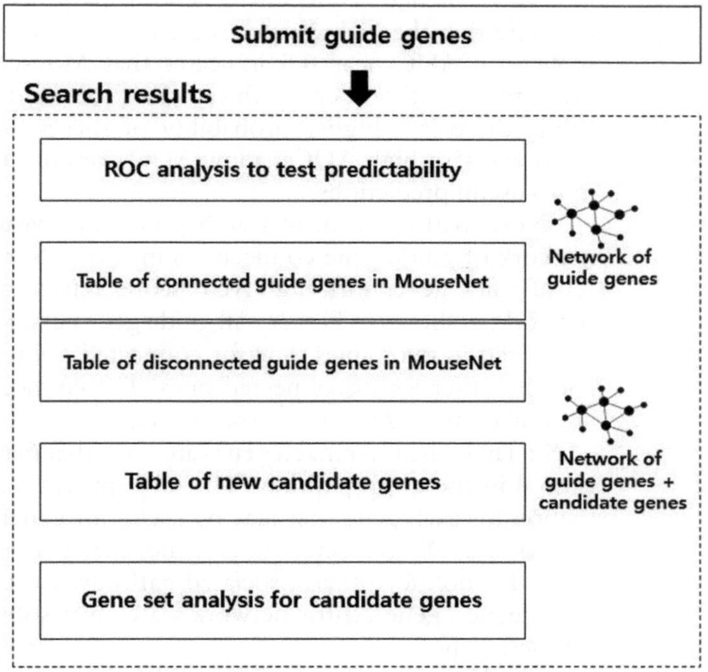

Fig. 1 Summary of pathway-centric network search. MouseNet performs network searches using user-submitted guide genes and returns various results: (1) ROC analysis for predictability test, (2) visualized network of guide genes, (3) table of connected guide genes, (4) table of disconnected guide genes, (5) visualized network of guide genes and new candidate genes, (6) table of new candidate genes. Lastly, the report page provides a tool for gene set analysis

represent a query pathway or phenotype. The MouseNet server provides searches for not only mouse but also eight other vertebrates through ortholog-based gene mapping provided by MGI [3]. If the target species is not mouse, the user should select the species name from the dropdown menu below the text box. Clicking the submission button then starts the network search. Search time is proportional to the number of guide genes and generally takes less than 30 s. Network searches return novel candidate genes as well as other information that helps users to generate functional hypotheses.

Prediction by guilt-by-association is expected to be more effective if the guide genes are well connected in the network. Accordingly, not every pathway is predictive in the network (*see* **Note 5**). To evaluate how predictive MouseNet is for a particular pathway (i.e., a set of guide genes), the network connectivity among guide genes is assessed based on receiver operating characteristic (ROC) analysis, which can be represented as the area under the ROC curve (AUC) score, ranging from 0.5 to 1 (Fig. 2a, b). A high AUC score means that each pathway gene (i.e., guide gene) tends to be retrieved in high ranks by network connections to other pathway member genes. It is generally accepted that AUC > 0.7 is an indication of high connectivity among pathway member genes, meaning that the pathway genes are highly predictive using MouseNet. An AUC near 0.5 indicates that MouseNet is no more predictive for the query pathway than the random expectation. While there is a higher probability of successful prediction for pathways with high AUCs, a low AUC does not necessarily result in no useful predictions.

Next, pathway-centric search produces a visualization of the network of guide gene connections in MouseNet (*see* **Note 6**). A set of guide genes with high AUC scores reflects a well-connected network, as shown in Fig. 2c. All guide gene network information is also listed in two tables, one for connected guide genes (Fig. 2d) and the other for disconnected ones. The connected guide genes are ranked from highest to lowest connectivity score (i.e., sum of LLS). Thus, the top ranked genes are hubs that may be functionally central in the given pathway. The table provides additional information for each gene that may be useful for generating functional hypotheses. Disconnected guide genes may belong to other pathways. To look for other associated pathways for the disconnected guide genes, gene-centric network search may subsequently be run on these genes.

The MouseNet server additionally returns a table of candidate genes ordered by connectivity (sum of LLS) to the guide genes (Fig. 3a). The table provides additional information that may help users in selecting final candidate genes for follow-up functional analyses. This includes information on component networks (*see* **Note 7**) that support the prediction, the list of linked guide genes,

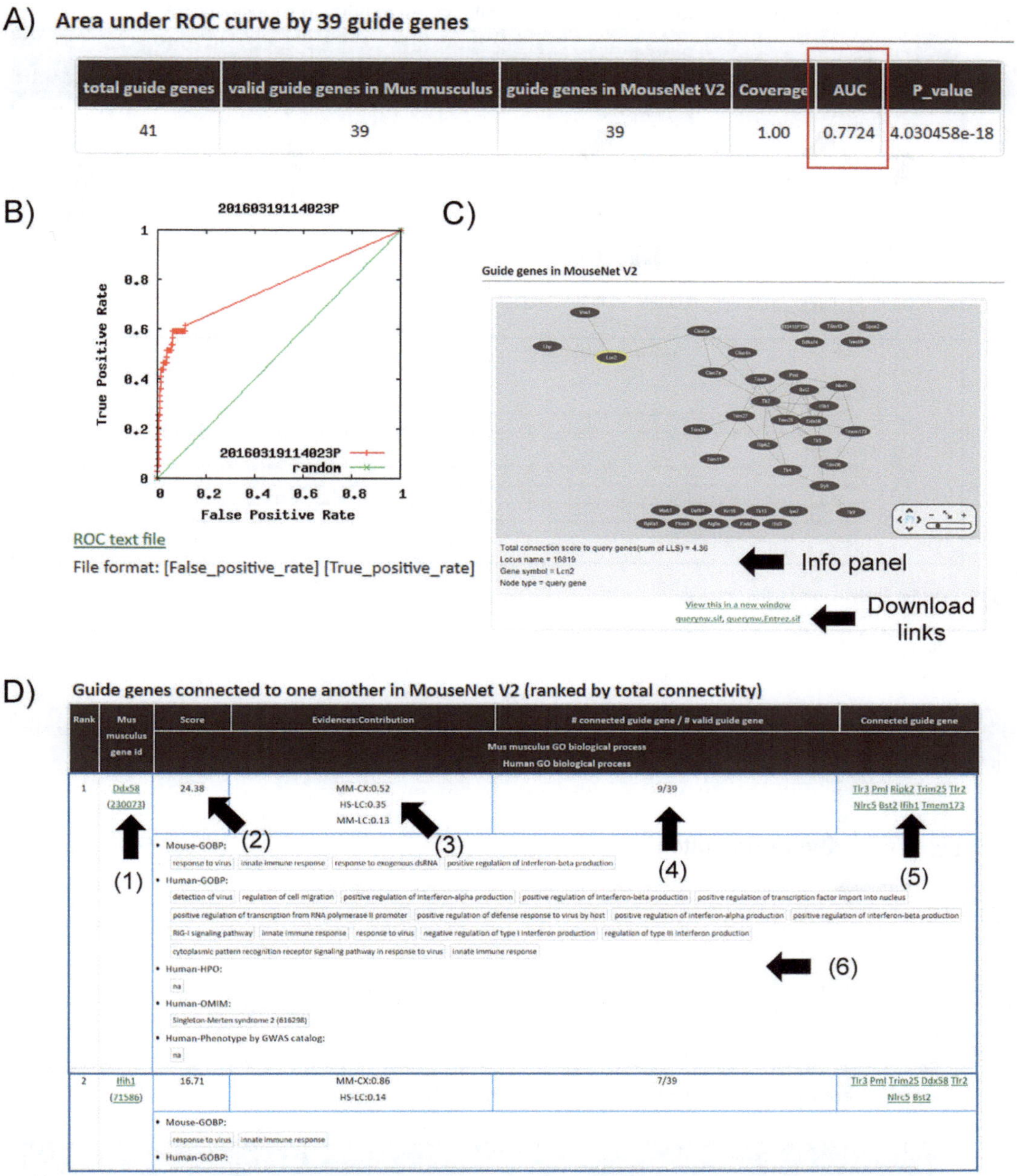

Fig. 2 Screenshot of the first half of the pathway-centric network search results of 41 guide genes for innate immune response. (**a**) Summary of ROC analysis of guide gene prediction. In this example, 39 of the 41 submitted genes are present in MouseNet, and their high predictability is indicated by the high AUC score (>0.77). (**b**) ROC curve plot. (**c**) Visualized network of guide genes. If the user clicks a node or an edge, detailed information appears in the "info panel" below. Users can also download network edge information using the "download links." (**d**) Table of connected guide genes sorted by total connectivity scores to other guide genes. Each guide gene comes with the following information: (*1*) gene ID (entrez ID), (*2*) score, (*3*) supporting component networks, (*4*) proportion of genes linked to the guide genes, (*5*) list of linked guide genes, and (*6*) functional annotations

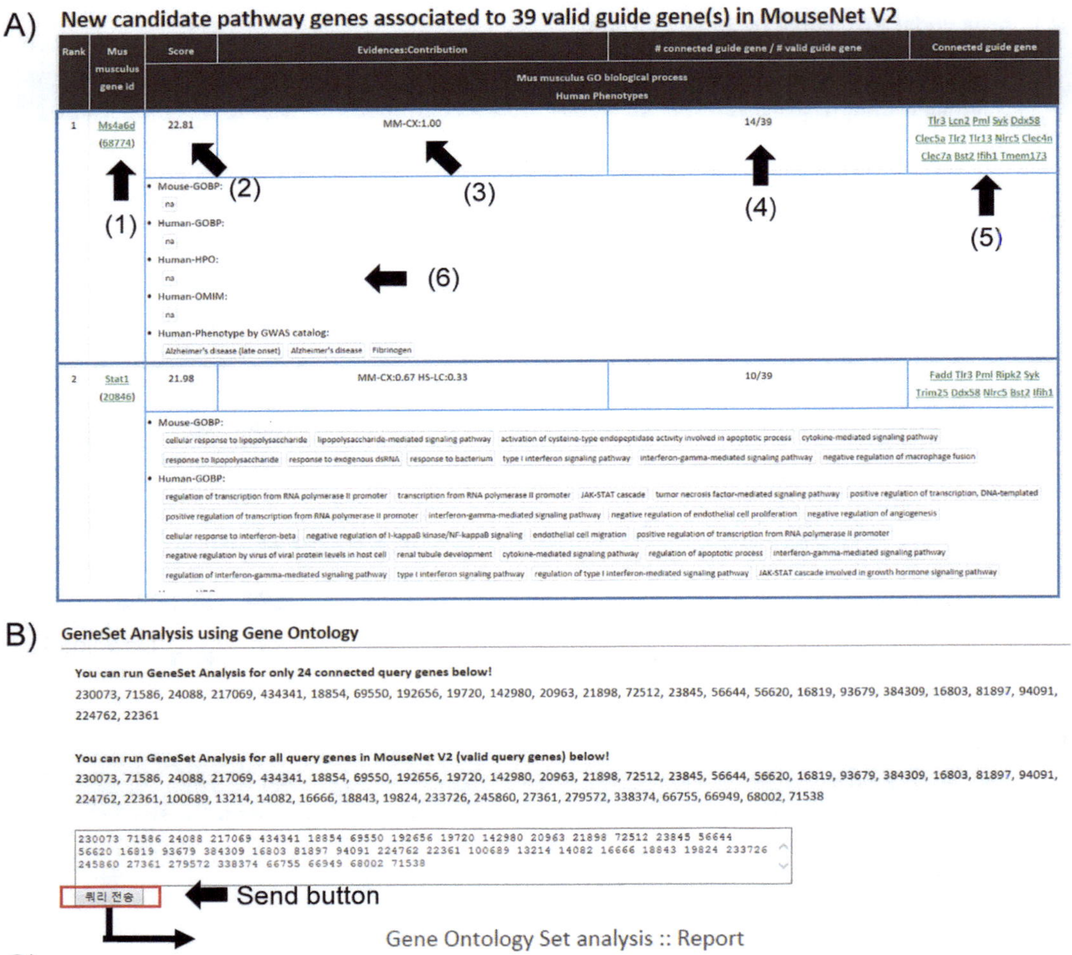

C)

Gene Ontology Set analysis :: Report

- Gene Ontology-Biological Process terms enrichment analysis result

If you want to see all of the results, see this report file: Report file

Description of columns: [Rank] [ID] [Description] [p-value (by Hypergeometric test)] [N = # of total genes] [m = # of query genes] [n = # of genes for the GO term] [k = # of genes for intersection between m and n]

Rank	ID	Description	p-value	N	m	n	k
1	GO:0045087	innate immune response	2.82e-119	20058	39	41	39
2	GO:0046597	negative regulation of viral entry into host cell	7.099e-18	20058	39	10	7
3	GO:1902187	negative regulation of viral release from host cell	3.61e-17	20058	39	6	6
4	GO:0032760	positive regulation of tumor necrosis factor production	6.821e-17	20058	39	24	8
5	GO:0009615	response to virus	3.517e-15	20058	39	37	8

Fig. 3 Screenshot of the second half of the pathway-centric network search results of 41 guide genes for innate immune response. (**a**) Table of novel candidate genes for innate immune response. All genes are sorted

and functional annotations. Lastly, users can predict associated GO annotations for the novel candidate genes by running a gene set enrichment test (Fig. 3b, c).

The search for novel pathway genes can be guided using genes annotated by various pathway databases, such as Gene ontology [2], KEGG [22], and MetaCyc [27]. If users want to identify novel genes involved in mouse phenotypes, genes annotated by the Mammalian Phenotype (MP) database [28] would provide appropriate guide genes. Due to the high phenotypic similarity between mouse and human, mouse disease genes would also serve as good guide genes in searches for novel human disease genes, or vice versa (*see* **Note 3**).

Another application of the pathway-centric search is the prioritization of genes from genetic screens, for example, if a set of genes for a phenotype is identified with a high-throughput forward screen and it is necessary to determine which genes are more functionally significant than others. One reasonable approach to such a problem would be to rank hits by total network connections to all other hits, assuming that hits of functional importance are connected to many other hits.

3.2 Predicting Pathways or Phenotypes for a Given Gene

The mouse genome is still largely unknown in terms of functions or associated phenotypes. A functional gene network can predict candidate functions of a query gene based on the functional annotations of its network neighbors. Since genes connected in a functional network are functionally coupled, functional gene information can be propagated to network neighbors. This idea is implemented in the MouseNet v2 server as the gene-centric network search. A gene-centric search starts with a submission of query genes and returns the top 30 predictions for six different functional and phenotypic annotations: Mouse-GOBP, Mouse-KEGG, Mouse-MP, Human-GOBP, Human-OMIM [23], and Human-HPO (human phenotype ontology) [29] (Fig. 4). The prediction score for each functional annotation term is calculated using the sum of LLS for all connected neighbors that are annotated with the term. Because scores are calculated for functional terms, a single neighboring gene may support multiple functional predictions. Human annotations are assigned to the mouse orthologs of human genes that provide functional annotations. While

Fig. 3 (continued) by their connectivity score to the guide genes. Each candidate gene comes with the information shown in Fig. 2d. Each candidate gene comes with the following information: (*1*) gene ID (entrez ID), (*2*) score, (*3*) supporting component networks, (*4*) proportion of genes linked to the guide genes, (*5*) list of linked guide genes, and (*6*) functional annotations. (**b**) Gene set analysis option to identify associated GO terms for novel candidate genes. Users can run the analysis by copying and pasting the new candidate genes into the text box and clicking the "send" button. (**c**) Results of gene set analysis. All GO terms with a p-value < 0.01 are listed in the table

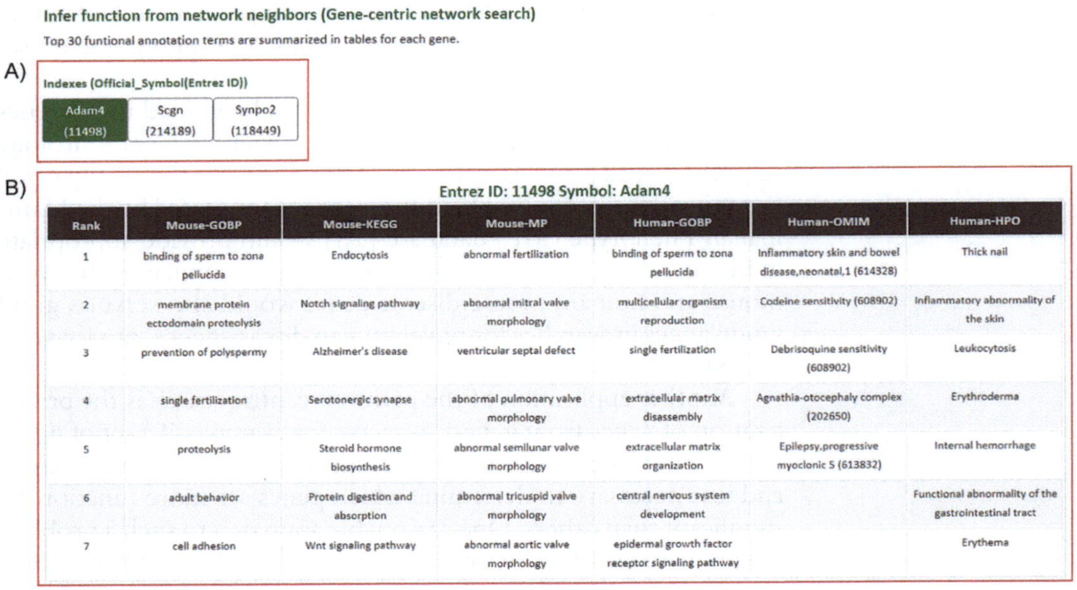

Infer function from network neighbors (Gene-centric network search)

Top 30 funtional annotation terms are summarized in tables for each gene.

A)

Indexes (Official_Symbol(Entrez ID))

Adam4 (11498) | Scgn (214189) | Synpo2 (118449)

B)

Entrez ID: 11498 Symbol: Adam4

Rank	Mouse-GOBP	Mouse-KEGG	Mouse-MP	Human-GOBP	Human-OMIM	Human-HPO
1	binding of sperm to zona pellucida	Endocytosis	abnormal fertilization	binding of sperm to zona pellucida	Inflammatory skin and bowel disease,neonatal,1 (614328)	Thick nail
2	membrane protein ectodomain proteolysis	Notch signaling pathway	abnormal mitral valve morphology	multicellular organism reproduction	Codeine sensitivity (608902)	Inflammatory abnormality of the skin
3	prevention of polyspermy	Alzheimer's disease	ventricular septal defect	single fertilization	Debrisoquine sensitivity (608902)	Leukocytosis
4	single fertilization	Serotonergic synapse	abnormal pulmonary valve morphology	extracellular matrix disassembly	Agnathia-otocephaly complex (202650)	Erythroderma
5	proteolysis	Steroid hormone biosynthesis	abnormal semilunar valve morphology	extracellular matrix organization	Epilepsy,progressive myoclonic 5 (613832)	Internal hemorrhage
6	adult behavior	Protein digestion and absorption	abnormal tricuspid valve morphology	central nervous system development		Functional abnormality of the gastrointestinal tract
7	cell adhesion	Wnt signaling pathway	abnormal aortic valve morphology	epidermal growth factor receptor signaling pathway		Erythema

Fig. 4 Screenshot of the gene-centric network search results of three query genes: *Adam4*, *Scgn*, and *Synpo2*. (**a**) Results for each submitted query gene can be viewed by selecting from among the gene selection boxes. (**b**) For each selected query gene, predictions from six functional/phenotypic annotations (Mouse-GOBP, Mouse-KEGG, Mouse-MP, Human-GOBP, Human-OMIM, and Human-HPO) are listed. Candidate annotation terms are ranked by sum of LLS to network neighbors annotated by the given annotation term

MouseNet provides up to 30 functional predictions from each annotation database for each gene, based on our experience, we recommend considering the top ten predictions for follow-up experimental validation.

To refine network search results, the MouseNet server also provides the option of filtering annotations by selected GOBP evidence codes. GO annotations include sources of supporting evidence, some of which are more reliable than others (*see* **Note 8**). By default, MouseNet conducts searches using annotations stemming from six reliable evidence codes based on experimental evidence and cited literature (IDA, IMP, IGI, IPI, IEP, and TAS). Including other evidence codes may generate more predictions but may result in lower accuracy.

3.3 Public Network Information as a Resource for Other Studies

The MouseNet web server provides a large amount of network information including various data types and organisms from the network-download page. Networks are downloadable with Entrez IDs or gene symbol names for laboratory mouse as well as the eight other model vertebrates in the system. Researchers may construct alternative versions of MouseNet using different methods to integrate component networks. Various co-expression networks derived from different experimental conditions would provide a useful resource for studying context-specific functional associations between genes and functions.

4 Case Studies

4.1 Predicting Novel Genes for Innate Immune Response

To identify new genes involved in innate immune response, we compiled 41 innate immune response genes annotated by the GOBP database. We conducted a pathway-centric network search by submitting the 41 genes as guide genes. From the returned search results, we found that the innate immune response pathway is highly predictive in MouseNet, with a high AUC score (AUC > 0.77) and a well-connected guide gene network (Fig. 2a–c). *Ms4a6d* was predicted as the top candidate gene for innate immune response (Fig. 3a). Interestingly, this gene is not annotated as an innate immune response gene by mouse GOBP, but its human ortholog was reported to be related to Alzheimer's disease, which is associated with innate immune response. Therefore, we consider *Ms4a6d* to be a strong candidate for a novel innate immune response gene. In addition, the second and third top candidate genes were annotated with "response to lipopolysaccharide" and "activation of innate immune response" terms, respectively. Because these terms are relevant to innate immunity, these genes are also strong candidates for innate immunity.

4.2 Predicting Novel Functions of Adam4, Scgn, and Synpo2

Functional annotation terms were predicted for three query genes that were not annotated by Mouse-GOBP by running gene-centric network searches (Fig. 4). The top candidate function for *Adam4* is "binding of sperm to zona pellucida," which was reported by Han et al. [30] based on experimental evidence. *Synpo2* was predicted to be involved in "retinal bipolar neuron differentiation" (first-ranked term), which were reported by Puthussery et al. [31]. Network search for *Scgn* predicted "muscle contraction" as second-ranked term, and the predicted function was reported by Turczyńska et al. [32]. These results demonstrate that MouseNet's gene-centric network search can recapitulate functional annotations that have been missed by human curators.

5 Notes

1. *Appropriate gene names for input*: Input genes for MouseNet searches must be Entrez IDs or official case-sensitive gene symbols that are separated by white space, new lines, or commas (e.g., "230,073, 71,586, 24,088, 217,069, 434,341" or "Ddx58, Ifih1, Tlr2, Trim25, Nlrc5"). Input data are case-sensitive because some vertebrate systems use gene symbols with the same spelling but different cases (e.g., uppercase for human genes but lowercase for zebrafish genes). Moreover, some vertebrate systems share gene symbols including the

case. Therefore, users must select the corresponding species for the input genes from the submission menus.

2. *Selection of guide genes from GOBP database with evidence codes*: GOBP annotations are easily downloadable as a gene2go file from the NCBI ftp server (ftp://ftp.ncbi.nlm.nih.gov/gene/DATA/gene2go.gz). To see, unzip the file (e.g., "gunzip gene2go.gz" in linux). The file contains tab-delimited information for mapping genes to GO terms. Each line contains eight columns: taxon ID, gene ID, GO ID, GO evidence code, qualifier, GO description, PubMed ID, and category. The taxon ID is used to select information for a specific target species (e.g., ~10,000 lines for mouse as of March 2016), and the category term "process" can be used to filter for GOBP annotations. In general, guide genes for BP terms with supporting evidence codes by experiments, such as IDA (inferred from direct assay), IMP (inferred from mutant phenotype), IGI (inferred from genetic interaction), IPI (inferred from physical interaction), IEP (inferred from expression pattern), and by literature citation with no experimental information (TAS, traceable author statement) provide more reliable seed information and better prediction results. However, if there are only a few genes annotated for a given GOBP term, one may need to consider including genes annotated by computational annotation with curation (e.g., ISS). We do not recommend using genes with electronic annotations with no curation (IEA) as guide genes.

3. *Selection of guide genes from disease gene databases*: OMIM is a major database for Mendelian diseases, whereas DO provides annotations for many complex diseases where a large number of genes make small contributions to pathogenicity. The GWAS catalog and GWASdb provide candidate disease genes from GWAS. Both databases provide probability scores of disease association for candidate genes. Users can thus select guide genes using their *p*-value threshold of choice. The GWAS catalog provides two alternative lists of candidate genes for the same GWAS analyses: candidates that are based on author reports and mapped candidates that are based on haplotype block data of disease-associated loci. We expect there may be a researcher's bias in the reported candidates. GWASdb contains significantly more GWAS analyses and, in turn, more candidate genes. One useful feature of this database is assignment of standard disease vocabulary by DO and HPO to each GWAS analysis, allowing for the compilation of a list of disease genes for each DO and HPO term.

4. *MouseNet edge weight by log likelihood score (LLS)*: MouseNet is a probabilistic functional network that provides the likelihood of functional coupling between connected genes as network edge weights. The LLS is calculated based on a Bayesian

statistics framework and is the ratio between posterior odds (i.e., updated belief by the given data) and prior odds (i.e., random expectation) that two genes belong to the same pathway.

5. *Low pathway or phenotype predictability*: There are two possible explanations for low predictability of a given pathway by MouseNet. First, MouseNet is not predictive for all pathways. Although the MouseNet model was trained to be general for any pathway rather than for specific ones, its predictive power may not be equally good for all pathways. Second, low predictability for phenotypes may occur because not all phenotypes are regulated by a specific pathway. For example, height and obesity are modulated by numerous genes that are involved in diverse pathways. If guide genes for a phenotype are derived from different pathways, they will not be well connected in MouseNet.
6. *Network visualization by MouseNet server or Cytoscape*: For network visualization in MouseNet server, Flash Player should be installed on the client machine. The network can be visualized and modified using the stand-alone software Cytoscape [33] and the sif files available from MouseNet search results. There are two options for downloading the network edge file for a network: nw.sif, which uses official gene symbols, and nw.Entrez.sif, which uses Entrez IDs. Cytoscape installation files can be downloaded from http://www.cytoscape.org.
7. *Component networks of MouseNet*: MouseNet was constructed by integration of 13 component networks inferred from various genomics data, network inference algorithms, and organisms. The component networks are coded as YY-XX, in which YY represents the organism (MM for mouse, DM for fly, HS for human, and SC for yeast) and XX represents the data type (CX for co-expression, GN for gene neighborhood, PG for phylogenetic profiling, LC for literature-based PPI, HT for high-throughput PPI, CC for co-citation, and GT for genetic interactions). All component networks are described in detail on the network-download page.
8. *GO evidence codes*: GO annotations from experimental or literature-based evidence tend to be more reliable than those from computational methods. Computational annotations can be improved with confirmation by curators. GO terms inferred from electronic annotations (IEA evidence code) are least reliable because they have not been confirmed by curators after automatic annotation. Detailed information about evidence codes is provided at http://geneontology.org/page/guide-go-evidence-codes.

Acknowledgments

This work was supported by grants from the National Research Foundation of Korea (2015R1A2A1A15055859, 2012M3A9B4028641, 2012M3A9C7050151).

References

1. Farrell CM, O'Leary NA, Harte RA, Loveland JE, Wilming LG, Wallin C, Diekhans M, Barrell D, Searle SM, Aken B, Hiatt SM, Frankish A, Suner MM, Rajput B, Steward CA, Brown GR, Bennett R, Murphy M, Wu W, Kay MP, Hart J, Rajan J, Weber J, Snow C, Riddick LD, Hunt T, Webb D, Thomas M, Tamez P, Rangwala SH, McGarvey KM, Pujar S, Shkeda A, Mudge JM, Gonzalez JM, Gilbert JG, Trevanion SJ, Baertsch R, Harrow JL, Hubbard T, Ostell JM, Haussler D, Pruitt KD (2014) Current status and new features of the Consensus Coding Sequence database. Nucleic Acids Res 42(Database issue):D865–D872. doi:10.1093/nar/gkt1059
2. Blake JA, Christie KR, Dolan ME, Drabkin HJ, Hill DP, Ni L, Sitnikov D, Burgess S, Buza T, Gresham C, McCarthy F, Pillai L, Wang H, Carbon S, Dietze H, Lewis SE, Mungall CJ, Munoz-Torres MC, Feuermann M, Gaudet P, Basu S, Chisholm RL, Dodson RJ, Fey P, Mi H, Thomas PD, Muruganujan A, Poudel S, Hu JC, Aleksander SA, McIntosh BK, Renfro DP, Siegele DA, Attrill H, Brown NH, Tweedie S, Lomax J, Osumi-Sutherland D, Parkinson H, Roncaglia P, Lovering RC, Talmud PJ, Humphries SE, Denny P, Campbell NH, Foulger RE, Chibucos MC, Giglio MG, Chang HY, Finn R, Fraser M, Mitchell A, Nuka G, Pesseat S, Sangrador A, Scheremetjew M, Young SY, Stephan R, Harris MA, Oliver SG, Rutherford K, Wood V, Bahler J, Lock A, Kersey PJ, McDowall MD, Staines DM, Dwinell M, Shimoyama M, Laulederkind S, Hayman GT, Wang SJ, Petri V, D'Eustachio P, Matthews L, Balakrishnan R, Binkley G, Cherry JM, Costanzo MC, Demeter J, Dwight SS, Engel SR, Hitz BC, Inglis DO, Lloyd P, Miyasato SR, Paskov K, Roe G, Simison M, Nash RS, Skrzypek MS, Weng S, Wong ED, Berardini TZ, Li D, Huala E, Argasinska J, Arighi C, Auchincloss A, Axelsen K, Argoud-Puy G, Bateman A, Bely B, Blatter MC, Bonilla C, Bougueleret L, Boutet E, Breuza L, Bridge A, Britto R, Casals C, Cibrian-Uhalte E, Coudert E, Cusin I, Duek-Roggli P, Estreicher A, Famiglietti L, Gane P, Garmiri P, Gos A, Gruaz-Gumowski N, Hatton-Ellis E, Hinz U, Hulo C, Huntley R, Jungo F, Keller G, Laiho K, Lemercier P, Lieberherr D, MacDougall A, Magrane M, Martin M, Masson P, Mutowo P, O'Donovan C, Pedruzzi I, Pichler K, Poggioli D, Poux S, Rivoire C, Roechert B, Sawford T, Schneider M, Shypitsyna A, Stutz A, Sundaram S, Tognolli M, Wu C, Xenarios I, Chan J, Kishore R, Sternberg PW, Van Auken K, Muller HM, Done J, Li Y, Howe D, Westerfield M, Consortium GO (2015) Gene Ontology Consortium: going forward. Nucleic Acids Res 43(D1): D1049–D1056. doi:10.1093/nar/gku1179
3. Blake JA, Bult CJ, Eppig JT, Kadin JA, Richardson JE, Grp MGD (2014) The Mouse Genome Database: integration of and access to knowledge about the laboratory mouse. Nucleic Acids Res 42(D1):D810–D817. doi:10.1093/nar/gkt1225
4. Shim JE, Lee I (2015) Network-assisted approaches for human disease research. Animal Cells Syst 19(4):231–235. doi:10.1080/19768354.2015.1074108
5. Cho A, Shin J, Hwang S, Kim C, Shim H, Kim H, Kim H, Lee I (2014) WormNet v3: a network-assisted hypothesis-generating server for *Caenorhabditis elegans*. Nucleic Acids Res 42(Web Server issue):W76–W82. doi:10.1093/nar/gku367
6. Shin J, Yang S, Kim E, Kim CY, Shim H, Cho A, Kim H, Hwang S, Shim JE, Lee I (2015) FlyNet: a versatile network prioritization server for the Drosophila community. Nucleic Acids Res 43(W1):W91–W97. doi:10.1093/nar/gkv453
7. Snider J, Kotlyar M, Saraon P, Yao Z, Jurisica I, Stagljar I (2015) Fundamentals of protein interaction network mapping. Mol Syst Biol 11(12):848. doi:10.15252/msb.20156351
8. Pellegrini M, Marcotte EM, Thompson MJ, Eisenberg D, Yeates TO (1999) Assigning protein functions by comparative genome analysis: protein phylogenetic profiles. Proc Natl Acad Sci U S A 96(8):4285–4288. doi:10.1073/pnas.96.8.4285
9. Overbeek R, Fonstein M, D'Souza M, Pusch GD, Maltsev N (1999) The use of gene clusters to infer functional coupling. Proc Natl Acad Sci

U S A 96(6):2896–2901. doi:10.1073/pnas.96.6.2896

10. Stuart JM, Segal E, Koller D, Kim SK (2003) A gene-coexpression network for global discovery of conserved genetic modules. Science 302 (5643):249–255. doi:10.1126/science.1087447
11. Kim E, Kim H, Lee I (2013) JiffyNet: a web-based instant protein network modeler for newly sequenced species. Nucleic Acids Res 41(W1):W192–W197. doi:10.1093/nar/gkt419
12. Lee I, Date SV, Adai AT, Marcotte EM (2004) A probabilistic functional network of yeast genes. Science 306(5701):1555–1558. doi:10.1126/science.1099511
13. Pena-Castillo L, Tasan M, Myers CL, Lee H, Joshi T, Zhang C, Guan YF, Leone M, Pagnani A, Kim WK, Krumpelman C, Tian WD, Obozinski G, Qi YJ, Mostafavi S, Lin GN, Berriz GF, Gibbons FD, Lanckriet G, Qiu J, Grant C, Barutcuoglu Z, Hill DP, Warde-Farley D, Grouios C, Ray D, Blake JA, Deng MH, Jordan MI, Noble WS, Morris Q, Klein-Seetharaman J, Bar-Joseph Z, Chen T, Sun FZ, Troyanskaya OG, Marcotte EM, Xu D, Hughes TR, Roth FP (2008) A critical assessment of *Mus musculus* gene function prediction using integrated genomic evidence. Genome Biol 9(Suppl 1):S2. doi:10.1186/Gb-2008-9-S1-S2
14. Kim WK, Krumpelman C, Marcotte EM (2008) Inferring mouse gene functions from genomic-scale data using a combined functional network/classification strategy. Genome Biol 9(Suppl 1):S5. doi:10.1186/Gb-2008-9-S1-S5
15. Barrett T, Wilhite SE, Ledoux P, Evangelista C, Kim IF, Tomashevsky M, Marshall KA, Phillippy KH, Sherman PM, Holko M, Yefanov A, Lee H, Zhang N, Robertson CL, Serova N, Davis S, Soboleva A (2013) NCBI GEO: archive for functional genomics data sets—update. Nucleic Acids Res 41(Database issue):D991–D995. doi:10.1093/nar/gks1193
16. Kim E, Hwang S, Kim H, Shim H, Kang B, Yang S, Shim JH, Shin SY, Marcotte EM, Lee I (2016) MouseNet v2: a database of gene networks for studying the laboratory mouse and eight other model vertebrates. Nucleic Acids Res 44(D1):D848–D854. doi:10.1093/nar/gkv1155
17. Kim H, Shin J, Kim E, Kim H, Hwang S, Shim JE, Lee I (2014) YeastNet v3: a public database of data-specific and integrated functional gene networks for *Saccharomyces cerevisiae*. Nucleic Acids Res 42(D1):D731–D736. doi:10.1093/nar/gkt981
18. Lee I, Blom UM, Wang PI, Shim JE, Marcotte EM (2011) Prioritizing candidate disease genes by network-based boosting of genome-wide association data. Genome Res 21 (7):1109–1121. doi:10.1101/gr.118992.110
19. Razick S, Magklaras G, Donaldson IM (2008) iRefIndex: a consolidated protein interaction database with provenance. BMC Bioinformatics 9:405. doi:10.1186/1471-2105-9-405
20. Shin J, Lee I (2015) Co-inheritance analysis within the domains of life substantially improves network inference by phylogenetic profiling. PLoS One 10(9):e0139006. doi:10.1371/journal.pone.0139006
21. Shin J, Lee T, Kim H, Lee I (2014) Complementarity between distance- and probability-based methods of gene neighbourhood identification for pathway reconstruction. Mol Biosyst 10(1):24–29. doi:10.1039/c3mb70366e
22. Kanehisa M, Goto S, Sato Y, Kawashima M, Furumichi M, Tanabe M (2014) Data, information, knowledge and principle: back to metabolism in KEGG. Nucleic Acids Res 42 (Database issue):D199–D205. doi:10.1093/nar/gkt1076
23. Amberger JS, Bocchini CA, Schiettecatte F, Scott AF, Hamosh A (2015) OMIM.org: Online Mendelian Inheritance in Man (OMIM(R)), an online catalog of human genes and genetic disorders. Nucleic Acids Res 43(Database issue):D789–D798. doi:10.1093/nar/gku1205
24. Kibbe WA, Arze C, Felix V, Mitraka E, Bolton E, Fu G, Mungall CJ, Binder JX, Malone J, Vasant D, Parkinson H, Schriml LM (2015) Disease Ontology 2015 update: an expanded and updated database of human diseases for linking biomedical knowledge through disease data. Nucleic Acids Res 43(D1):D1071–D1078. doi:10.1093/nar/gku1011
25. Welter D, MacArthur J, Morales J, Burdett T, Hall P, Junkins H, Klemm A, Flicek P, Manolio T, Hindorff L, Parkinson H (2014) The NHGRI GWAS Catalog, a curated resource of SNP-trait associations. Nucleic Acids Res 42 (D1):D1001–D1006. doi:10.1093/nar/gkt1229
26. Li MJ, Liu Z, Wang P, Wong MP, Nelson MR, Kocher JP, Yeager M, Sham PC, Chanock SJ, Xia Z, Wang J (2016) GWASdb v2: an update database for human genetic variants identified by genome-wide association studies. Nucleic Acids Res 44(D1):D869–D876. doi:10.1093/nar/gkv1317
27. Caspi R, Altman T, Billington R, Dreher K, Foerster H, Fulcher CA, Holland TA, Keseler IM, Kothari A, Kubo A, Krummenacker M, Latendresse M, Mueller LA, Ong Q, Paley S,

Subhraveti P, Weaver DS, Weerasinghe D, Zhang PF, Karp PD (2014) The MetaCyc database of metabolic pathways and enzymes and the BioCyc collection of pathway/genome databases. Nucleic Acids Res 42(D1): D459–D471. doi:10.1093/nar/gkt1103

28. Smith CL, Eppig JT (2012) The mammalian phenotype ontology as a unifying standard for experimental and high-throughput phenotyping data. Mamm Genome 23(9-10):653–668. doi:10.1007/s00335-012-9421-3

29. Kohler S, Doelken SC, Mungall CJ, Bauer S, Firth HV, Bailleul-Forestier I, Black GCM, Brown DL, Brudno M, Campbell J, FitzPatrick DR, Eppig JT, Jackson AP, Freson K, Girdea M, Helbig I, Hurst JA, Jahn J, Jackson LG, Kelly AM, Ledbetter DH, Mansour S, Martin CL, Moss C, Mumford A, Ouwehand WH, Park SM, Riggs ER, Scott RH, Sisodiya S, Van Vooren S, Wapner RJ, Wilkie AOM, Wright CF, Vulto-van Silfhout AT, de Leeuw N, de Vries BBA, Washingthon NL, Smith CL, Westerfield M, Schofield P, Ruef BJ, Gkoutos GV, Haendel M, Smedley D, Lewis SE, Robinson PN (2014) The Human Phenotype Ontology project: linking molecular biology and disease through phenotype data. Nucleic Acids Res 42(D1):D966–D974. doi:10.1093/nar/gkt1026

30. Han C, Choi E, Park I, Lee B, Jin S, Kim do H, Lee B, Nishimura H, Cho C (2009) Comprehensive analysis of reproductive ADAMs: relationship of ADAM4 and ADAM6 with an ADAM complex required for fertilization in mice. Biol Reprod 80(5):1001–1008. doi:10.1095/biolreprod.108.073700

31. Puthussery T, Gayet-Primo J, Taylor WR (2010) Localization of the calcium-binding protein secretagogin in cone bipolar cells of the mammalian retina. J Comp Neurol 518(4):513–525. doi:10.1002/cne.22234

32. Turczynska KM, Sward K, Hien TT, Wohlfahrt J, Mattisson IY, Ekman M, Nilsson J, Sjogren J, Murugesan V, Hultgardh-Nilsson A, Cidad P, Hellstrand P, Perez-Garcia MT, Albinsson S (2015) Regulation of smooth muscle dystrophin and synaptopodin 2 expression by actin polymerization and vascular injury. Arterioscler Thromb Vasc Biol 35(6):1489–1497. doi:10.1161/ATVBAHA.114.305065

33. Shannon P, Markiel A, Ozier O, Baliga NS, Wang JT, Ramage D, Amin N, Schwikowski B, Ideker T (2003) Cytoscape: a software environment for integrated models of biomolecular interaction networks. Genome Res 13(11):2498–2504. doi:10.1101/gr.1239303

Chapter 15

The FANTOM5 Computation Ecosystem: Genomic Information Hub for Promoters and Active Enhancers

Imad Abugessaisa, Shuhei Noguchi, Piero Carninci, and Takeya Kasukawa

Abstract

The Functional Annotation of the Mammalian Genome 5 (FANTOM5) project conducted transcriptome analysis of various mammalian cell types and provided a comprehensive resource to understand transcriptome and transcriptional regulation in individual cellular states encoded in the genome.

FANTOM5 used cap analysis of gene expression (CAGE) with single-molecule sequencing to map transcription start sites (TSS) and measured their expression in a diverse range of samples. The main results from FANTOM5 were published as a promoter-level mammalian expression atlas and an atlas of active enhancers across human cell types. The FANTOM5 dataset is composed of raw experimental data and the results of bioinformatics analyses. In this chapter, we give a detailed description of the content of the FANTOM5 dataset and elaborate on different computing applications developed to publish the data and enable reproducibility and discovery of new findings. We present use cases in which the FANTOM5 dataset has been reused, leading to new findings.

Keywords FANTOM5, Genomics, Transcriptome, Genome annotation, CAGE, TSS, Promoter, Enhancer

1 Introduction

To deeply study and understand the basic mechanisms of living systems and their cellular functions at the molecular level, scientists are identifying and quantifying the abundance of different types of RNA molecules in single-cells or bulk cell populations [1]. Mapping transcriptional start sites (TSSs) of the RNA and the promoter landscape in the genome is crucial for understanding gene regulation by transcription factors and active enhancers [2].

The need for profiling the genome-wide TSS led to the development of a cap analysis of gene expression (CAGE). CAGE is a high-throughput method intended to capture and sequence the 5′-end of transcripts and hence enable detailed mapping of TSS regions [3–5].

Daisuke Kihara (ed.), *Protein Function Prediction: Methods and Protocols*, Methods in Molecular Biology, vol. 1611, DOI 10.1007/978-1-4939-7015-5_15, © Springer Science+Business Media LLC 2017

CAGE enabled the discovery of promoters (their location) with their expression level as well as active enhancers in mammalian genomes [6]. CAGE protocols have been utilized in high-profile research projects such as FANTOM3 [7], ENCODE [8], modENCODE [9] and for mapping zebrafish core promoters [6].

The RIKEN-led FANTOM consortium has contributed to the advancement of mammalian genome research. Scientists from around the globe work together to produce, analyze, interpret, and publish genomic data with annotations for the research community [10]. This effort has resulted in a comprehensive database resource of mammalian gene expression across human and mouse genomes.

In the FANTOM5 project, we mapped TSS and profiled promoter activities of various mammalian cells. FANTOM5, conducted in two phases, used ~1800 human and ~1000 mouse samples. In phase 1, FANTOM5 published the promoter atlas of various human and mouse cells and tissues [11] and an atlas of human active enhancers [12]. The main publications are complemented with a set of satellite ones. The promoter atlas shows that mammalian promoters comprise multiple TSSs; these TSSs are expressed differently in different samples. The atlas of active enhancers reveals cell and tissue specificity of the expression of the active enhancers. In phase 2 of FANTOM5, the focus has been on a panel of 19 human and 14 mouse time course samples (sampling the transition between different cell types or developmental stages), applying the same sequencing methods as in phase 1 (single-molecule sequencing), and extended the total number of core promoters in human and mouse to 201,802 and 158,966 respectively [13]. As with promoters, the set of transcribed enhancers was extended to 65,423 in human and 44,459 in mouse.

The datasets published by the FANTOM5 consortium consist of three types. The first contains the primary analyses, consisting of genomic information like the definition of TSS regions and their association with known genes. The second contains the outcome of the advanced analyses. The third is the metadata about the FANTOM5 samples and experiments (*see* **Note 1**).

For external users to handle the above datasets, several computing applications have been developed. Their functions complement each other and form an ecosystem for sharing and publishing the FANTOM5 data [14]. The web interfaces for FANTOM5 data resources, software, protocols, publications, and other relevant materials are accessible at http://fantom.gsc.riken.jp/5/.

2 Materials

This section discusses the software applications that have been developed in FANTOM5. Fig. 1 illustrates the FANTOM5

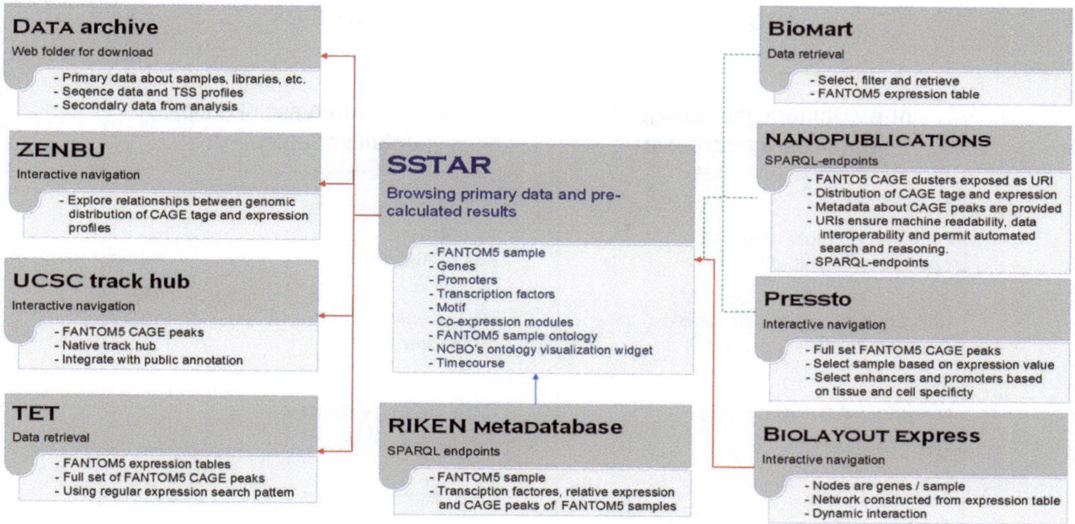

Fig. 1 FANTOM5 computational ecosystem. Most of the applications are connected directly to SSTAR (*red lines*) or accessible directly independent of SSTAR (*green dashed line*). The applications are developed to allow interactive navigation and data query and retrieval for any subset of the data. Part of the SSTAR data are deposited in RIKEN metadatabase (*blue line*)

software applications with their data content, accessing methods, types of user interface, and relationships to other applications. These applications support some of the data exploration, visual inspection, querying, and download of the dataset of interest.

2.1 FANTOM5 Data Archive

The main experimental and analysis data obtained in FANTOM5 have been assembled together and formatted using data standards. These data were made available for bulk download from the Internet through the FANTOM5 file browser (see http://fantom.gsc.riken.jp/5/data/search/?fd=latest). As shown in Fig. 2, the FANTOM5 dataset is divided into two parts, the basic dataset and the extra dataset. The logic behind this division is that the primary data related to metadata of the samples, RNA and sequencing libraries, and sequences (with their genome mapping and CAGE tag starting site (CTSS) files) reside together as a basic dataset from the experimental protocol (*see* **Note 2** about CTSS). The secondary bioinformatics analyses are collected and stored under the extra dataset folder.

2.1.1 FANTOM5 Basic Data Archive

The basic dataset folder holds the primary results from the FANTOM5 project pipeline described in [14]. The basic folder is further divided into subfolders; each folder is named according to a simple and unified schema. Each folder name consists of three parts:

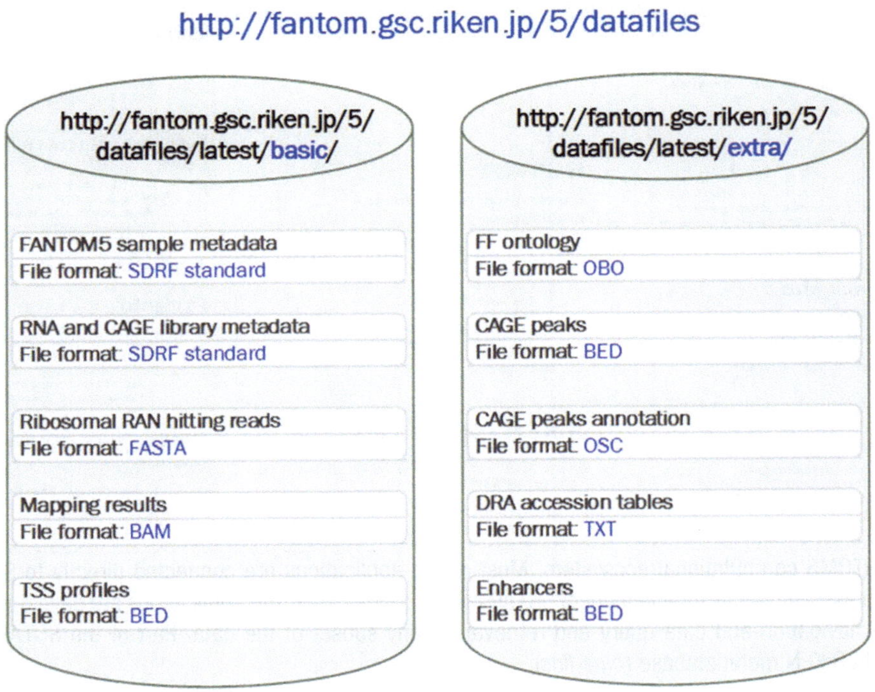

Fig. 2 FANTOM5 Data Archive content. Data Archive are divided into two parts, the basic data folder and the extra data folder

1. Human or mouse,
2. Biological category (i.e., primary cell, cell line, time course, etc.), and
3. The sequencing technology used, e.g., HeliscopeCAGE sequencing.

An example of a folder is human.cell_line.hCAGE. In the human.cell_line.hCAGE folder, sequences, genome mapping, and CTSS profiles of all human cell-line samples that have been sequenced with HeliscopeCAGE sequencers are downloadable (*see* **Note 3**). As illustrated in Fig. 2, several types of files are stored using standard formats such as sample and data relationship format (SDRF) for sample RNA and sequencing library information [15], FASTA format for sequences [16], BAM format for mapped genomic coordination, and BED format for CTSS profiles. For further reading about the sample naming schema used in FANTOM5 project, see [14].

2.1.2 FANTOM5 Extra Data Archive

The extra dataset folder holds several results of the secondary analysis from the basic dataset (*see* Fig. 2). The CAGE peak folder contains genomic coordination of the CAGE peaks with their annotations and expression profiles. The CAGE peaks were generated with the decomposition-based peak identification (DPI) method (*see* (11) and **Note 2**), which aims to identify units of TSS regions that could be controlled by the same regulation. As a result, two subsets of the CAGE peaks have been released: the robust set and the permissive set. They were obtained based on tag evidence thresholds. The permissive threshold is that a peak has a single CTSS supported by three or more observations in at least one profile, which corresponds to a ratio of promoter to exonic peaks of ~0.7. While the robust threshold is that a peak has a CTSS supported by 11 or more observations and 1 or more TPM, which corresponds to a ratio of promoter/exonic of ~2.0 [11]. The total number of CAGE peaks identified by the permissive threshold is 1,048,124 in human and 652,860 in mice. While the defined robust CAGE peaks are 184,827 and 116,277 for human and mouse respectively. The difference in total number of CAGE peaks between the permissive and robust sets is due to the fact that the robust set used a more stringent threshold compared to the permissive. As mentioned in [11], several subsequence analyses performed in FANTOM5 to define human and mouse promoters used the robust set only while the permissive set was used to identify the active enhancers.

The second folder in the extra dataset is the list of tables with DRA accession numbers of sequences deposited in the DNA Data Bank of Japan (DDBJ) Sequence Read Archive (DRA) [17]. Additional annotations of the CAGE peaks are provided in the CAGE peaks annotation folder. CAGE peak enhancers from FANTOM5 (phases 1 and 2) are provided in the folder Enhancers. The enhancer folder mainly contains a list of the sample names in text format, as well as expression counts and expression tables for those active enhancers. The last folder in the extra archive is the FANTOM5 ontology (FF ontology) file in Open Biomedical Ontologies (OBO) format, FF ontology merged the standard cell ontology [18], disease [19], and anatomy ontology [20] with FANTOM5 sample classifications. FF ontology is used to annotate the FANTOM5 panel of samples to increase discoverability and sharing across different communities. For further understanding of the structure in the extra dataset folder, *see* **Note 3**.

2.2 FANTOM5 Web Applications

In FANTOM5 we realized that different users have different interests and use various scenarios when accessing the comprehensive data resources provided by FANTOM5. The users' interests vary from browsing and exploring the dataset to find their sample or gene of interest, to downloading an expression table or genomic view of a CAGE peak. Advanced users are more interested in downloading a subset of the data to reproduce the results and try to develop and test their new hypotheses. Considering this mixture of interests and different user profiles (e.g., biologists and bioinformaticians), we developed several web-based systems. The systems are interconnected and provide an easy-to-use interface for users. All the systems provide basic functionalities such as browsing, querying, extracting, and exporting subsets of the data.

The FANTOM5 consortium released the entire datasets under the Creative Commons license, and interfaces to the web application are open for use without a prior registration or password.

Below this, we will explain each of the applications developed, with their content and functions.

2.2.1 ZENBU (http://fantom.gsc.riken.jp/zenbu/)

A genome browser to explore and visualize genomic and transcriptomic data is indispensable for genomic research. Considering this requirement, the ZENBU genome browser was developed [21]. ZENBU provides all functions for handling the vast amount of samples from FANTOM5 with the ability to browse RNA-Seq and CAGE data. Among other features, ZENBU enables direct upload of BAM files and provides a secure environment for collaboration and data sharing. The system has embedded data processing and analysis capabilities and is flexible enough to be configured in an intuitive and dynamic way. The main interface of the ZENBU genome browser consists of a set of menu buttons providing browsing, data exploration, and user data upload and sharing. The last button is for user documentation and help.

FANTOM5 data in ZENBU are stored and illustrated as a set of ZENBU views. Each view is dedicated to a certain type of FANTOM5 dataset. The preconfigured views are

1. FANTOM5 Human promoterome with gene expression
2. FANTOM5 Human time course
3. FANTOM5 Mouse promoterome, and
4. FANTOM5 Mouse time course

The ZENBU genome browser hosts several public annotation tracks for overlaying with FANTOM5 CAGE data.

2.2.2 SSTAR (http://fantom.gsc.riken.jp/5/sstar/Main_Page)

The FANTOM5 dataset is heterogeneous due to the large number and diversity of the sample types used during the project. Adding to the heterogeneity of the raw data, the huge size and the divergence of the analysis results would make it difficult to access and discover the FANTOM5 dataset. To solve this issue, we have developed a web-based system called the Semantic Catalogue of Samples, Transcription initiation and Regulators (SSTAR) [22]. The system utilizes Semantic MediaWiki (SMW) as a core component for building the application [23]. SSTAR stores all samples (human and mouse) annotated with FF ontology, and the system provides access to the integrated sample metadata and analysis results from the Sample page. Another category of the data stored in SSTAR is the transcription factors and co-expression modules discovered in FANTOM5. In SSTAR, many data are displayed in embedded tabular format and graphical layout. This is to allow visual selection and inspection of subsets of the displayed data.

We have embedded a view of the UCSC genome browser in SSTAR, and the system is hyperlinked to several external databases, such as DDBJ, HGNC, MIM, Ensemble, HPRD, and Vega. From SSTAR the user can also access other FANTOM5 web applications via hyperlinks. Additionally, in SSTAR we embedded the NCBO's Ontology visualization widget for viewing the FF ontology tree.

2.2.3 FANTOM5's BioMart Instance (http://fantom.gsc.riken.jp/5/biomart/)

Aiming at providing the sequenced transcription start site regions, the FANTOM5 consortium installed an instance of the open source BioMart [24]. Advanced settings in BioMart permit integration with Bioconductor via web service and API interface. The current installation of the BioMart instance holds the dataset from FANTOM5 phase 1.1 and phase 2.0. The dataset collection consists of samples and CAGE peak annotation for human and mouse. This simple interface enables data filtering using combo boxes and checkboxes, complemented with free text search capabilities. The search result is shown in a text format that can be exported in different formats.

2.2.4 TET (http://fantom.gsc.riken.jp/5/tet)

Another tool for querying the FANTOM5 dataset is TET (table extraction tool). TET stores expression tables that are available in the tab-separated value format and compressed for download from the FANTOM5 archive data. TET provides a search function for the user with the capability to query and extract a subset of the FANTOM5 expression table based on the name of the sample and the name of the CAGE peak. Users of the TET are able to select a set of annotation attributes to be added to the selected expression table (e.g., short_description, hgnc_id, uniport_id, etc.). The result of the online query can either be shown as a table or downloaded for reuse. The dataset available in TET is of FANTOM5 phase 1.1 and phase 2.0.

2.2.5 Biolayout Express (http://fantom.gsc.riken.jp/5/biolayout/?cultureKey=&q=5/biolayout)

The applications discussed so far provide access points to the FANTOM5 dataset in different formats for download and reuse. Meanwhile, for quickly understanding a large-scale dataset, visualization tools have played important roles. One of the recent developments in biological data visualization is the use of the network paradigm, to illustrate the relationship between different biological entities (gene and protein samples). The Biolayout is one such application that uses the network concept to enable biological date integration, visualization, and analysis [25]. FANTOM5 has utilized Biolayout features for part of the data analysis in the project.

The data input for Biolayout is in expression format (co-expression graph), and the output is a directional network graph, which has already been precomputed in FANTOM5. The node in the resulting network represents a FANTOM5 sample or promoter and the edge shows similarly expressed samples or promoters. The resulting network enables an identification of network patterns in the dataset. All sample nodes in the Biolayout are connected to SSTAR. Clicking any of them, the user can find a sample page in SSTAR.

2.2.6 PrESSTo, the FANTOM5 Promoter and Enhancer Slider Tool (http://pressto.binf.ku.dk/)

Another application interface for the FANTOM5 dataset is the PrESSTo tool. This tool enables selecting a promoter or enhancer that is specific to a cell type or a tissue. PrESSTo uses the expression tables from FANTOM5 as measured in tags per million (TPM), and the samples are aggregated in a group of facets. Each facet includes a set of samples that are related to each other anatomically or functionally over the whole human body. PrESSTo consists of two applications. The first is the Human Promoter Expression Atlas, the second is the Human Transcribed Enhancer Atlas. The main feature of PrESSTo is enabling the identification of promoters or enhancers simply by using a slider to dynamically change the expression constraints. The resulting set of promoters or enhancers can be exported for further analysis outside PrESSTo.

2.2.7 Nanopublication (http://fantom5.nanopub.org/sparql)

The FANTOM5 consortium utilized the development in semantic technologies for publishing the dataset from the project. Semantic technologies have features to provide a resource over the web in machine readable and discoverable format. The semantic web technology uses the Resource Description Framework (RDF) as a formal language for describing the resource [26]. The resource could be a dataset or web service. Nanopublication as an application of the semantic web is defined as an interoperable exchange publication format that can describe each data point, e.g., a CAGE peak as assertion and provide detailed information about the assertion as metadata. RDF is used to serialize a set of statements that compose the Nanopublication. Each Nanopublication is then defined by using resolvable uniform resource identifiers (URIs) [27]. Three types of Nanopublication have been introduced for the FANTOM5 dataset: individual CAGE peaks, their associated genes, and the expression values in TPM for each gene [14].

The SPARQL (SPARQL Protocol and RDF Query Language) endpoints are hosted by Database Center for Life Science in JAPAN and deployed using the Virtuoso SPARQL Query Editor. The published SPARQL endpoints are accessible at http://fantom5.nanopub.org/sparql.

2.2.8 RIKEN Metadatabase (http://metadb.riken.jp/metadb/db/SciNetS_ria365i)

The RIKEN metadatabase is a collection of metadata resources for life science database generated by research groups in RIKEN. Similar to Nanopublication, the RIKEN metadatabase employs semantic web technology and provides a set of the SPARQL-endpoints.

Part of the FANTOM5 dataset from SSTAR is hosted by RIKEN metadatabase, and metadata about the FANTOM5 sample, CAGE profiles, expressions, etc. are searchable and downloadable via a graphical user interface at http://metadb.riken.jp/metadb/db/SciNetS_ria365i.

Currently, FANTOM5 data in the RIKEN metadatabase consist of several classes: human and mouse samples, transcription factors, relative expression of human and mouse samples, and CAGE peaks of human and mouse samples. The statistics about the FANTOM5 database in the RIKEN metadatabase are provided at http://metadb.riken.jp/.

2.2.9 UCSC FANTOM5 Data Hub

UCSC (The University of California Santa Cruz) genome browser is widely used in genomic research [28]. Preconfigured data files from FANTOM5 (CAGE peaks and TSS profiles) are available via track hub in the UCSC genome browser. This gives users means to integrate and overlay FANTOM5 data with annotation databases and other native track hubs maintained in the UCSC genome browser. To access the FANTOM5 track hub, a user is required to access the UCSC genome browser home at www.genome.ucsc.edu, then import FANTOM5 hub for phase 1 and phase 2. The track hub is processed for human in hg38 and hg19 assemblies and for mouse in mm10 and mm9.

3 Methods

Here, we illustrate different methods and techniques to use the listed applications in the materials section. This will be followed with a set of use cases on how to perform specific tasks.

3.1 Harvesting FANTOM5 Biological Samples

All the samples used in FANTOM5 are listed in SSTAR. Each sample is assigned a unique sample ID, e.g., 10000-101A1. Each sample has a page that could be shared (e.g., http://fantom.gsc.riken.jp/5/sstar/FF:10000-101A1). Samples are classified based on manual curation and FF ontology. Table 1 shows types and number of sample collection per species in FANTOM5.

The typical usage is to search for a sample related to your research, see if this sample has been measured in FANTOM5, and find enriched transcription factor(s) in the specific sample.

To perform this kind of a task, the user performs the following steps (*see* Fig. 3):

1. Access SSTAR at http://fantom.gsc.riken.jp/5/sstar/Main_Page.
2. Select a species, i.e., human or mouse sample.

Table 1
Statistics of FANTOM5 sample collection

Category Species	Primary cells	Cell lines	Development	Tissues	Time courses	Control
Human	561	268	0	183	783	21
Mouse	140	0	257	26	591	4

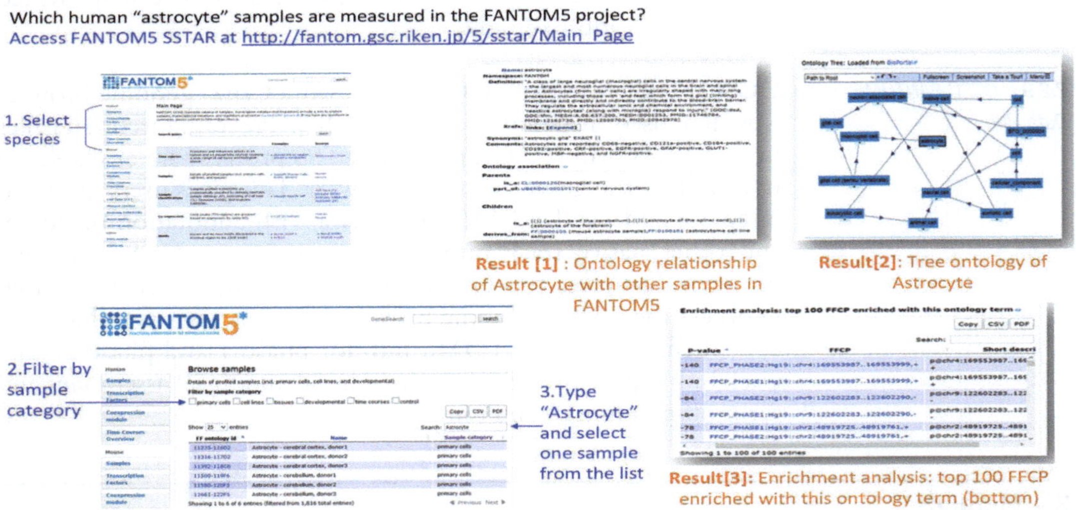

Fig. 3 Search for FANTOM5 sample(s). SSTAR user interface allows users to search for a particular sample or group of samples of interest

3. Filter by a sample category as in Table 1. This enables you to find a group of samples based on manually curated classifications.
4. Search by a sample keyword on a filtered list of samples.
5. Select one sample from the list of samples.

3.1.1 Result of the Above Steps

The resulting sample page provides the following information:

1. Sample metadata details.
2. Genomic view of the sample in the UCSC genome browser and ZENBU.
3. Sample and library accession numbers.
4. Transcription factors with enriched expression in the sample.

5. Co-expression clusters with enriched expression in the sample.
6. Repeat families with enriched expression in the sample.
7. TFBS (DNA) motifs overrepresented in proximal region of promoters active in this sample.
8. FANTOM5 phase 1 novel unique motifs.
9. De novo motifs identified by HOMER in promoters active in this sample.
10. FF ontology parent-child relationships in an ontology tree view.

Another approach to search for FANTOM5 samples is to launch BioMart at http://fantom.gsc.riken.jp/5/biomart/ and type the name of the sample.

3.1.2 Explore Similarities Between Sample and Transcription Initiation Activity

To find and explore similarities between FANTOM5 samples or between transcription initiation activities, Biolayout Express is an appropriate tool that allows performing the analysis in an interactive way using the concept of network. The followings are the bases for using Biolayout Express:

1. Network nodes represent either samples or CAGE peaks.
2. Node colors are based on the co-expression cluster they belong to.
3. Network edges represent correlation between two nodes (samples).

To find similarity relationships between samples, carry out the following steps:

1. Launch Biolayout Express at http://fantom.gsc.riken.jp/5/biolayout/.
2. Select one of the following Java Web Start files for download or display on the fly
   (a) Human primary cells.
   (b) Promoters for transcription factors only.
   (c) A sample-to-sample correlation matrix.
   (d) Mouse cell and tissues.
   (e) Promoters for transcription factors only.
   (f) A sample-to-sample correlation matrix.
   (g) Co-expression clustering of all human promoters in FANTOM5.
3. Search for samples or CAGE peaks and connect to either SSTAR or ZENBU.

Fig. 4 Search for a highly expressed gene in FANTOM5 set of samples. SSTAR users are able to search for gene by symbol, description, or ID

4. Find groups (clusters) of samples sharing similar patterns in expression. This is possible by performing on-the-fly clustering of the nodes.
5. Export the result of the view for further use.

3.1.3 Find highly Expressed Gene and Regulators

One of the potential uses of the biological database is to find a gene of interest and know the set of samples in which the gene is highly expressed. This feature is supported by a simple and easy-to-use interface in SSTAR. The following steps will help in finding genes that are highly expressed in a set of FANTOM5 samples (*see* Fig. 4).

1. Access SSTAR at http://fantom.gsc.riken.jp/5/sstar/Main_Page.
2. On the top right menu, in the search box called Gene Search, type EntrezGene ID, symbol, or any free texts, e.g., aaRS.
3. If any genes are found, users can display each gene page with detailed information about the genes.

Result of the above steps

1. List of gene attributes [Symbol: AARS, Description: alanyl-tRNA synthetase; Synonyms: CMT2N; Species: Human (*Homo sapiens*)].

2. Set of links to external resources to get more information about the gene, e.g., HGNC, MIM, Ensembl, HPRD and Vega.
3. If available, associated JASPAR motifs and Transcription factors.
4. Links to a list of FANTOM CAGE peaks (TSS regions) in which the gene is associated. Each link will provide access to detailed information on enrichment analysis and the co-expression cluster the gene belongs to.
5. Genomic view of the gene in the UCSC genome browser.
6. Table with a list of FANTOM5 samples in which the gene is expressed or active, with the expression values in TPM.
7. ENCODE TF ChIP-Seq peak enrichment analysis (if available).

3.1.4 Retrieve an Expression Table for a Specific Sample Set

The purpose here is to search and download the expression table for a set of FANTOM5 samples. This is possible to perform by using the TET tool. The following steps are required, as demonstrated in Fig. 5:

1. Launch FANTOM5 TET at http://fantom.gsc.riken.jp/5/tet/.
2. Choose "Expression (RLE normalized) . . ." from the dataset list box.
3. Click the column list box. From there you can filter samples of interest.

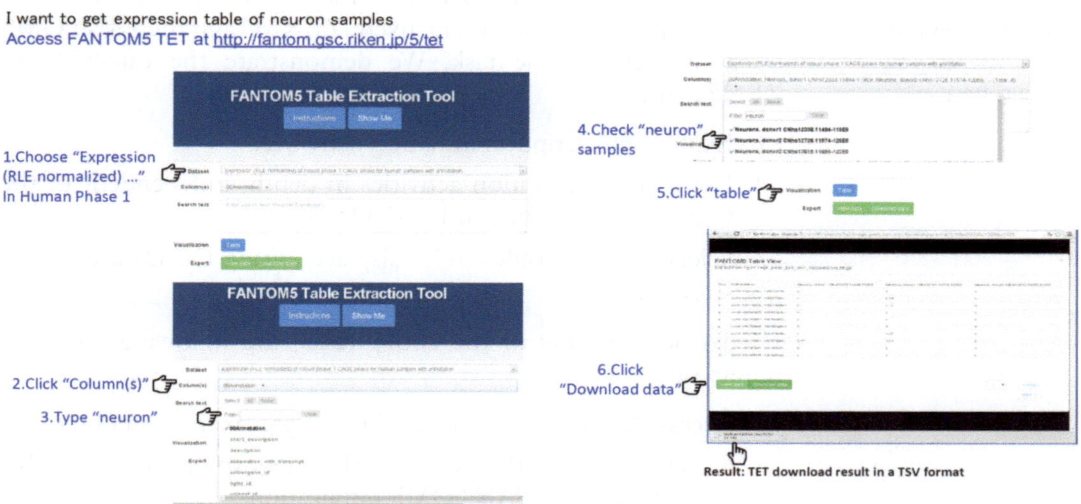

Fig. 5 Download of FANTOM5 expression table from TET tool. FANTOM5 expression tables are downloadable from TET application. TET enables the retrieval of expression table for a particular sample from the FANTON5 panel of samples

4. Check samples that you are interested in.
5. From visualization, select Table.
6. Select Download Data.

3.1.5 Fetch a List of Promoters in a Specific Chromosome

Some researchers have special interests in a specific genomic region or chromosome. The researcher might be interested in finding identified promoters in the chromosome of interest, such as chr21 in the human genome. Through the following steps, a list of promoters detected in FANTOM5 within chr21 in hg19 can be retrieved and downloaded:

1. Launch FANTOM5 BioMart at http://fantom.gsc.riken.jp/5/biomart/.
2. Start a new MartView session.
3. Select a dataset from the list menu on the left-hand side, and pick the dataset.
4. From the left menu, select filter and then check CAGE peaks ID. In the same view, type "chr21:%."
5. From the left-hand side, check the attributes you are interested in. Several attributes are available for selection under CAGE peaks, annotations, GENCODE annotation, and repeat annotation.
6. Select Results on the top menu; this will display a table with a list of annotations for download.

3.1.6 Navigating FANTOM5 Sample and CAGE Peak Using the Genome Browser

To navigate and explore transcription activities and their annotation on the genomic browser interactively, ZENBU and the FANTOM5 track hub enable the task. We demonstrate the capability of ZENBU to:

1. Explore transcription initiation activities.
2. Examine transcription activities in depth, independent of the CAGE peaks defined in FANTOM5.
3. Select CAGE profiles to be displayed using the Data Explorer search tab.

To perform any of the above tasks, the following steps are required:

1. Access ZENBU from http://fantom.gsc.riken.jp/zenbu/.
2. Select All FANTOM5 Views in the data explorer to get a table of all views from FANTOM5 publications.
3. Alternatively, select any of the preconfigured pooled tracks of the FANTOM5 CAGE profiles and search for your gene/sample of interest. The result will be a view of the accumulated transcription activities. Users are able to filter the view by sample names and sort the result by expression levels.

3.2 Case Studies

In this section, we provide interesting case studies using FANTOM5 datasets, and reported to interesting biological findings.

3.2.1 Case Study 1

"Partitioning heritability by functional annotation using genome-wide association summary statistics" [29].

This article introduced a new computation method, the stratified LD score regression. The method partitions heritability from genome-wide association study (GWAS) summary statistics. Using this method, the authors analyzed a set of functional elements to estimate their polygenic contributions to heritability in GWAS. Using the stratified LD methods, they apply it to 17 complex diseases and traits. They started by analyzing non-cell-type specific annotations and defined heritability enrichment in many of these functions. Their approach allows them to identify a large immunological disease-specific enrichment of heritability in FANTOM5 enhancers.

3.2.2 Case Study 2

"Transcriptome analysis of recurrently deregulated genes across multiple cancers identifies new pan-cancer biomarkers" [30].

This study aims to identify a set of deregulated genes in cancer to be used as biomarkers and therapeutic targets. The authors compared expressions of primary cells and cancer cell lines using RNA-Seq samples from the Cancer Genome Atlas and the FANTOM5 dataset. The results of this analysis reveal a core transcript set and a set of enhancer RNAs upregulated in cancer, as well as detecting a set of promoters overrepresenting repetitive elements that are often upregulated in cancer.

4 Notes

1. In FANTOM5, a collection of samples from human and mouse primary cells (in steady state) has been complemented with profiles of different cancer cell lines representing distinct cancer subtypes, 152 human post-mortem tissues and 271 mouse developmental tissue samples.

   Basic dataset

   The sample information was annotated using FANTOM5 ontology (FF ontology), which merged several types of standard ontologies. In addition to sample metadata, for each sample the mapped sequence data in BAM formats and the profile TSS in BED format were released.

Dataset from primary analysis

The second type of dataset was the primary analysis, consisting of genomic information like the definition of TSS regions and their association with known genes. This is considered primary analysis; the resulting data is used for further advanced analysis.

Dataset from secondary analysis

The third type of data released by FANTOM5 is the outcome of the advanced analysis. This kind of data includes co-expression clustering of TSSs, statistical assessment of transcription factor binding motifs within CAGE peaks, samples, and enrichment analysis of pathways or samples.

2. CAGE transcription start sites defined in http://fantom.gsc.riken.jp/5/datahub/hg38/TSS_activity.html, the CAGE reads from the sequenced 5′-ends of the transcript are counted. The counts are performed at a single base pair resolution on the genomic coordinates. This represents TSS activities in the sample. CAGE peaks across FANTOM5 samples are identified by decomposition-based peak identification (DPI). Each of the peaks consists of neighboring and related TSSs.
3. The data generated during the FANTOM5 project has been published in different ways. However, the entire dataset is available as data archive and easy to access via file browser at http://fantom.gsc.riken.jp/5/data/. The file is available for download using one of the following methods:
   (a) Use LFTP command: lftp http://fantom.gsc.riken.jp/5/datafiles/latest/.
   (b) Use http download server: http://fantom.gsc.riken.jp/5/datafiles/latest.
   (c) Use FANTOM5 file browser: http://fantom.gsc.riken.jp/5/data/.

   The archive is divided into main two folders, the basic and the extra folder. Each of the main folders is organized in two or three levels.

   Level 1:

   Basic level listing a group of folders.

   http://fantom.gsc.riken.jp/5/data/search/?fd=phase2.0%2Fbasic

   Level 2:

   List several folders and each of the folders contains a list of files. For example, human cell line samples that were processed with HeliscopeCAGE sequencing are available at:

   http://fantom.gsc.riken.jp/5/datafiles/latest/basic/human.cell_line.LQhCAGE/

- **BASIC**
  - • human.cell_line.hCAGE
    - 00_human.cell_line.hCAGE.hg19.assay_**sdrf.txt**
    - B lymphoblastoid cell line: GM12878 ENCODE, biol_rep3.CNhs12333.10823-111C4.hg19.nobarcode.**bam**
    - B lymphoblastoid cell line: GM12878 ENCODE, biol_rep2.CNhs12332.10822-111C3.hg19.nobarcode.rdna.**fa.gz**
    - Burkitt's lymphoma cell line:RAJI.CNhs11268.10476-106I8.hg19.ctss.**bed.gz**
  - human.cell_line.LQhCAGE
  - human.fractionation.hCAGE
  - human.tissue.hCAGE
  - mouse.tissue.LQhCAGE
- mouse.tissue.LQhCAGE

- **EXTRA**
  - • CAGE_peaks
    - hg19.cage_peak_phase1and2combined_ann.**txt.gz**
    - hg19.cage_peak_phase1and2combined_coord.**bed.gz**
    - hg19.cage_peak_phase1and2combined_counts.**osc.txt.gz**
  - DRA_accession_tables
  - CAGE_peaks_annotation
  - Enhancers
  - Ontology

Fig. 6 FANTOM5 data archive and its sub-folders. The data archive is divided into two folders, the basic and the extra folder. The basic folder stores the primary data while the extra folder stores the secondary bioinformatics analysis

Level 3:

List of files for a particular sample, the first file contains all metadata about the human cell lines and is stored in SDRF format. Other files such as mapped sequence files and TSS profiles are also available (*see* Fig. 6).

Acknowledgment

FANTOM5 was made possible by a Research Grant for RIKEN Omics Science Center from MEXT to Yoshihide Hayashizaki and a Grant of the Innovative Cell Biology by Innovative Technology (Cell Innovation Program) from the MEXT, Japan to Yoshihide Hayashizaki and to the RIKEN Center for Life Science Technologies. This study is also supported by Research Grants from the Japanese Ministry of Education, Culture, Sports, Science and Technology through RIKEN Centre for Life Science Technologies, Division of Genomic Technologies. We would like to thank all the members of the FANTOM5 consortium for contributing to the generation of samples and analysis of the data-set and thank GeNAS for data production.

References

1. Wang Z, Gerstein M, Snyder M (2009) RNA-Seq: a revolutionary tool for transcriptomics. Nat Rev Genet 10(1):57–63. doi:10.1038/nrg2484
2. Yu NY, Hallstrom BM, Fagerberg L et al (2015) Complementing tissue characterization by integrating transcriptome profiling from the human protein atlas and from the FANTOM5 consortium. Nucleic Acids Res 43 (14):6787–6798. doi:10.1093/nar/gkv608
3. Murata M, Nishiyori-Sueki H, Kojima-Ishiyama M et al (2014) Detecting expressed genes using CAGE. In: Miyamoto-Sato E, Ohashi H, Sasaki H, Nishikawa J-I, Yanagawa H (eds) Transcription factor regulatory networks: methods and protocols. Springer, New York, NY, pp 67–85. doi:10.1007/978-1-4939-0805-9_7
4. Kanamori-Katayama M, Itoh M, Kawaji H et al (2011) Unamplified cap analysis of gene expression on a single-molecule sequencer. Genome Res 21(7):1150–1159. doi:10.1101/gr.115469.110
5. Takahashi H, Lassmann T, Murata M et al (2012) 5[prime] end-centered expression profiling using cap-analysis gene expression and next-generation sequencing. Nat Protoc 7(3):542–561
6. Haberle V, Li N, Hadzhiev Y et al (2014) Two independent transcription initiation codes overlap on vertebrate core promoters. Nature 507(7492):381–385. doi:10.1038/nature12974
7. Carninci P, Kasukawa T, Katayama S et al (2005) The transcriptional landscape of the mammalian genome. Science 309 (5740):1559–1563. doi:10.1126/science.1112014
8. The ENCODE project consurtium (2012) An integrated encyclopedia of DNA elements in the human genome. Nature 489(7414):57–74
9. Celniker SE, Dillon LAL, Gerstein MB et al (2009) Unlocking the secrets of the genome. Nature 459(7249):927–930
10. De Hoon M, Shin JW, Carninci P (2015) Paradigm shifts in genomics through the FANTOM projects. Mamm Genome 26 (9–10):391–402. doi:10.1007/s00335-015-9593-8
11. Forrest AR, Kawaji H, Rehli M et al (2014) A promoter-level mammalian expression atlas. Nature 507(7493):462–470. doi:10.1038/nature13182
12. Andersson R, Gebhard C, Miguel-Escalada I et al (2014) An atlas of active enhancers across human cell types and tissues. Nature 507 (7493):455–461. doi:10.1038/nature12787
13. Arner E, Daub CO, Vitting-Seerup K et al (2015) Gene regulation. Transcribed enhancers lead waves of coordinated transcription in transitioning mammalian cells. Science 347 (6225):1010–1014. doi:10.1126/science.1259418
14. Lizio M, Harshbarger J, Shimoji H et al (2015) Gateways to the FANTOM5 promoter level mammalian expression atlas. Genome Biol 16:22. doi:10.1186/s13059-014-0560-6
15. The Functional Genomics Data Society (FGED), TFGDS MAGE-TAB Specification. http://www.mged.org/mage-tab/
16. Pearson WR, Lipman DJ (1988) Improved tools for biological sequence comparison. Proc Natl Acad Sci U S A 85(8):2444–2448
17. Kodama Y, Mashima J, Kaminuma E et al (2012) The DNA data Bank of Japan launches a new resource, the DDBJ Omics archive of functional genomics experiments. Nucleic Acids Res 40(Database issue):D38–D42. doi:10.1093/nar/gkr994
18. Bard J, Rhee SY, Ashburner M (2005) An ontology for cell types. Genome Biol 6(2): R21. doi:10.1186/gb-2005-6-2-r21
19. Kibbe WA, Arze C, Felix V et al (2015) Disease ontology 2015 update: an expanded and updated database of human diseases for linking biomedical knowledge through disease data. Nucleic Acids Res 43(Database issue): D1071–D1078. doi:10.1093/nar/gku1011
20. Druzinsky R, Mungall C, Haendel M et al (2013) What is an anatomy ontology? Anat Rec (Hoboken) 296(12):1797–1799. doi:10.1002/ar.22805
21. Severin J, Lizio M, Harshbarger J et al (2014) Interactive visualization and analysis of large-scale sequencing datasets using ZENBU. Nat Biotechnol 32(3):217–219. doi:10.1038/nbt.2840
22. Abugessaisa I, Shimoji H, Sahin S et al (2016) FANTOM5 transcriptome catalog of cellular states based on Semantic MediaWiki. Database (Oxford) 2016. doi:10.1093/database/baw105
23. Krötzsch M, Vrandečić D, Völkel M (2006) Semantic MediaWiki. In: Cruz I, Decker S, Allemang D et al (eds) The semantic web-ISWC 2006, Lecture notes in computer science, vol 4273. Springer, Berlin Heidelberg, pp 935–942. doi:10.1007/11926078_68
24. Smedley D, Haider S, Durinck S et al (2015) The BioMart community portal: an innovative

alternative to large, centralized data repositories. Nucleic Acids Res 43(W1): W589–W598. doi:10.1093/nar/gkv350

25. Theocharidis A, van Dongen S, Enright AJ et al (2009) Network visualization and analysis of gene expression data using BioLayout express (3D). Nat Protoc 4(10):1535–1550. doi:10.1038/nprot.2009.177
26. Abugessaisa I (2010) Geospatial metadata extraction from product description document applying methods from ontology engineering. Int J Metadata Semant Ontologies 5 (4):321–332. doi:10.1504/ijmso.2010.035554
27. Patrinos GP, Cooper DN, van Mulligen E et al (2012) Microattribution and nanopublication as means to incentivize the placement of human genome variation data into the public domain. Hum Mutat 33(11):1503–1512. doi:10.1002/humu.22144
28. Speir ML, Zweig AS, Rosenbloom KR et al (2016) The UCSC genome browser database: 2016 update. Nucleic Acids Res 44(D1): D717–D725. doi:10.1093/nar/gkv1275
29. Finucane HK, Bulik-Sullivan B, Gusev A et al (2015) Partitioning heritability by functional annotation using genome-wide association summary statistics. Nat Genet 47(11): 1228–1235
30. Kaczkowski B, Tanaka Y, Kawaji H et al (2016) Transcriptome analysis of recurrently deregulated genes across multiple cancers identifies new pan-cancer biomarkers. Cancer Res 76 (2):216–226. doi:10.1158/0008-5472.CAN-15-0484

Chapter 16

Multi-Algorithm Particle Simulations with Spatiocyte

Satya N.V. Arjunan and Koichi Takahashi

Abstract

As quantitative biologists get more measurements of spatially regulated systems such as cell division and polarization, simulation of reaction and diffusion of proteins using the data is becoming increasingly relevant to uncover the mechanisms underlying the systems. Spatiocyte is a lattice-based stochastic particle simulator for biochemical reaction and diffusion processes. Simulations can be performed at single molecule and compartment spatial scales simultaneously. Molecules can diffuse and react in 1D (filament), 2D (membrane), and 3D (cytosol) compartments. The implications of crowded regions in the cell can be investigated because each diffusing molecule has spatial dimensions. Spatiocyte adopts multi-algorithm and multi-timescale frameworks to simulate models that simultaneously employ deterministic, stochastic, and particle reaction-diffusion algorithms. Comparison of light microscopy images to simulation snapshots is supported by Spatiocyte microscopy visualization and molecule tagging features. Spatiocyte is open-source software and is freely available at http://spatiocyte.org.

Keywords Mathematical modeling, Biophysical simulation, Biochemical simulation, Particle simulation, Multi-algorithm simulation, Diffusion

1 Introduction

Heterogeneity and complex behavior observed at the cellular scale can arise from basic reaction-diffusion interactions at the molecular scale. Cell polarization, morphogenesis, chemotaxis, and cytokinesis are some of the processes strongly coupled with the diffusion and spatiotemporal localization of signaling proteins. In addition to noisy reactions from low reactant numbers, the stochastic nature of the molecular interactions can also induce oscillations [1] and excite systems [2] in space and time. Consequently, spatial stochastic simulators have become important tools to elucidate the molecular mechanisms driving these processes [3–5]. Spatial simulators are also necessary for bottom-up construction of biophysically and biochemically realistic whole cell models [6]. Current simulators can be broadly categorized as (1) mesoscale, with the space discretized into subvolumes [7–12]; (2) microscale, with each molecule tracked individually [13–19]; and (3) hybridized meso- and

Daisuke Kihara (ed.), *Protein Function Prediction: Methods and Protocols*, Methods in Molecular Biology, vol. 1611,
DOI 10.1007/978-1-4939-7015-5_16, © Springer Science+Business Media LLC 2017

microscale [20–22]. Mesoscopic simulators can advance time faster than microscale methods because they only track the concentration of species in the subvolumes instead of the exact position of each molecule. However, with the absence of such positional information, mesoscopic simulators cannot depict the effects of individual molecular interactions such as molecular crowding [17, 19] and rebinding [15].

Spatiocyte is a hybrid macro- and microscale simulator for stochastic reaction-diffusion processes. Molecule species homogeneously distributed (HD) in a compartment are simulated at the macroscopic scale, whereas heterogeneously distributed (non-HD) molecules are individually diffused in the lattice. Species found in many copies, which are usually evenly distributed in a compartment, can be simulated rapidly at the macroscopic scale by only tracking their molecule number without explicit diffusion steps. Spatiocyte can simulate reaction-diffusion of molecules on filaments such as microtubules, in addition to membranes and in solutions [23]. The simulator includes a visualizer that can display the position of non-HD molecules and their time-averaged trajectory, as visualized using a light microscope, for rapid visual comparison to experimentally captured images. Compartment geometry can be specified using a combination of geometric primitives such as cuboids, ellipsoids, and cylinders, which can be translated and rotated. Spatiocyte can virtually tag and track a subpopulation of molecules individually, even as they transition from one state to another or between compartments. With this feature, it is possible to compare the diffusion behavior of the molecules with sparsely tagged molecules in experiments. Metabolic reactions usually involve a large number of molecules, most of which can be assumed to be HD. These reactions can be executed using Michaelis–Menten, Gillespie, or mass action algorithms alongside Spatiocyte reaction-diffusion processes.

A detailed description of the Spatiocyte particle simulation algorithm is provided by Arjunan and Tomita [24]. Briefly, the space is discretized into hexagonal close-packed (HCP) lattice with regular sphere voxels. Each voxel can be occupied by a single non-HD molecule as the voxel size approximates the size of the molecule. A diffusing non-HD molecule can walk to one of its 12 neighbor voxels by random selection in a diffusion step interval. The interval is calculated from the species diffusion coefficient and the voxel radius. The walk is successful if the target voxel is vacant; otherwise, if it contains a reactant pair, a collision occurs and the molecules react with a probability corresponding to the reaction rate constant. If the voxel is occupied by a nonreactive molecule, the source molecule stays in the original voxel. HD molecules react with non-HD molecules in an event-driven manner according to a spatially adapted Next Reaction method [24, 25]. The diffusion and reaction processes of the simulation method have been verified with analytical and numerical solutions [24].

Using Spatiocyte and protein measurements such as concentration, binding, dissociation, and diffusion constants in different states (monomer, homodimer, heterodimer, cytosolic, membrane-bound, etc.), we can build a quantitative model that closely represents the cellular system of interest. We can verify the model by comparing simulation outcomes with wild-type and mutant phenotypes of the system. Often, not all model parameters can be measured and in such cases, the parameters can be estimated by adjusting them until the simulated phenotypes agree with observations. Since Spatiocyte simulations are rapid, we can use the model to explore the parameter space to predict the system behavior with different combinations of parameter values. For example, we can predict the phenotype with varying expression and activity levels of proteins (e.g., phosphorylation and dephosphorylation), protein mutation (removal of one or more of its functions), cell morphology, and initial conditions. These predictions can be tested experimentally. If what we observe in experiments do not match the predictions, the model can be adjusted until it recapitulates the observations. The adjustments will provide new insights about the molecular mechanisms underlying the system. In a more advanced case, Spatiocyte can also be used to design models, made up of components with well-characterized parameters, to generate a particular behavior of a system (e.g., spatial pattern or domain formation). The model can then be realized experimentally to generate such behaviors.

2 Materials

2.1 Input Model Parameters

Spatiocyte requires simulation model parameters to write the model and perform simulations. The model parameters include the simulation voxel size, compartment geometry and dimensions, molecule species, species molecule number and diffusion coefficient, reactions and their rate constants. In the Methods section, the steps to build a multi-algorithm particle simulation model are provided. There are also several other example models included in the software package.

2.2 Spatiocyte Software

Spatiocyte runs stably on Ubuntu Linux but it is still experimental on Mac OS X and Windows systems. Up-to-date download and installation instructions of Spatiocyte can be found at http://spatiocyte.org. On a fresh Ubuntu system, Spatiocyte requires additional libraries and packages to run. These include Python, Boost.Python, Git, Hierarchical Data Format 5, Matplotlib, NumPy, SciPy, and GNU Scientific Library (GSL), all of which are automatically installed along with Spatiocyte. After the model is parsed, the simulator creates the compartments according to the specified geometry and populates the molecules as given in the initial conditions. Logger modules are also initialized and they start logging simulation data in the specified log files. After that, graphical or command line interface is activated to transfer the control of program execution to the user. The user can run the simulation for a given time and view the dynamics with the Spatiocyte Visualizer. The visualizer is a separate program in the Spatiocyte package that can run concurrently with the simulator. It loads the log file created by the visualization logger modules to display time-lapse molecule positions or simulated microscopy snapshots using the OpenGL library. Screenshots of the simulation can be saved in Portable Network Graphics (PNG) format and animated to create high-resolution movies.

3 Methods

3.1 Build a Simulation Model

Here, we will build a multi-algorithm simulation model in Python. The model can also be written C++ but for simplicity, we focus on Python (*see* **Note 1**). The algorithm modules, such as diffusion, reaction, logger, tagger, and molecule population, can be specified as necessary in a model. Each module has its own set of options that are defined in the model by the user. These options are described in the general guide to build Spatiocyte models [26]. Our multi-algorithm model consists of mass action, Spatiocyte next-reaction, and lattice-based particle reaction-diffusion methods.

1. *Prepare model parameters.* The parameters of the model are listed in Fig. 1. In the model, we have non-HD species, A, B, and C, with molecules individually diffused, and HD species, E, S, ES, and P. We use mass action to simulate three reversible Michaelis–Menten-type reactions, where a product P is formed by a single enzyme-substrate complex, ES from a single substrate, S and an enzyme, E: $S + E \underset{k_2}{\overset{k_1}{\rightleftharpoons}} ES \overset{k_3}{\rightharpoonup} E + P$. The product P will stochastically bind A to generate a heterodimer B with Spatiocyte next-reaction method: $P + A \overset{k4}{\rightharpoonup} B$. The fifth and final reaction, which is diffusion-influenced, involves reactant B that will bind another diffusing reactant A to generate C: $A + B \overset{k5}{\rightharpoonup} C$.
2. *Open an empty model file.* To write the model, open an editor and save an empty file as ode-snrp-particle.py. Add the Python code as provided in the steps below to the file and save as we go along. The complete model file is included in the examples/ode-snrp-particle directory of the Spatiocyte package.
3. *Specify steppers and voxel radius.* In the model, we refer to the Spatiocyte simulator as `theSimulator`. Since the term is used

Voxel radius	4.4 nm

Compartment	Geometry	Length X	Length Y	Length Z
root	Cuboid	500 nm	500 nm	500 nm

Species	Type	Initial number	Diffusion coefficient
A	non-HD	1500	0.0005 $\mu m^2 s^{-1}$
B	non-HD	0	0.0005 $\mu m^2 s^{-1}$
C	non-HD	0	0.0005 $\mu m^2 s^{-1}$
E	HD	100	Not applicable
S	HD	1000	Not applicable
ES	HD	0	Not applicable
P	HD	0	Not applicable

Reaction	Reaction method	Type	Rate constant
E + S → ES	Mass action	Deterministic	1e-22 $m^3 s^{-1}$
ES → E + S	Mass action	Deterministic	0.1 s^{-1}
ES → E + P	Mass action	Deterministic	0.1 s^{-1}
P + A → B	Spatiocyte next-reaction	Stochastic	5e-24 $m^3 s^{-1}$
A + B → C	Diffusion-influenced	Stochastic	5e-24 $m^3 s^{-1}$

Fig. 1 Input parameters of the multi-algorithm simulation model

frequently throughout the code, we define `sim` as a short alias for `theSimulator` below. All Spatiocyte models require the SpatiocyteStepper. It advances the time steps of logger, reaction, and diffusion modules in an event-driven manner. We set the radius of the hexagonal close-packed lattice voxels with the VoxelRadius option (*see* **Note 2**). Below, we have set the radius to 4.4 nm. Since we are building a multi-algorithm model consisting of mass action reactions, we also need to specify a stepper for ordinary differential equation solver called ODEStepper to execute the reactions in continuous time. The remaining two reactions, next-reaction and diffusion-influenced-reaction, are performed in discrete time by the SpatiocyteStepper. We set the maximum step interval of the ODEStepper to 0.001 s for accuracy (*see* **Note 3**). Each compartment in the Spatiocyte model must have its StepperID assigned to the SpatiocyteStepper ID. This model only consists of the root compartment, rootSystem and we set its StepperID to SpatiocyteStepper id, 'SS' (*see* **Note 4**).

```
sim = theSimulator
s = sim.createStepper('SpatiocyteStepper', 'SS')
s.VoxelRadius = 4.4e-9
s = sim.createStepper('ODEStepper', 'DE')
s.MaxStepInterval = 1e-3
sim.rootSystem.StepperID = 'SS'
```

4. *Set compartment geometry.* We can specify the geometry of the root compartment by setting the GEOMETRY variable to one of the six supported geometric primitives: cuboid ('0'), ellipsoid ('1'), cylinder ('2'), rod ('3'), pyramid ('4') and erythrocyte ('5'). For simplicity, we use a cuboid compartment geometry in this model. More complicated geometries can be constructed using a combination of the existing primitives. The three variables LENGTH[X, Y, Z] specify the compartment lengths in the direction of [x, y, z]-axis, respectively. Here, the lengths are set to 500 nm.

```
sim.createEntity('Variable', 'Variable:/:GEOMETRY').Value
= 0
sim.createEntity('Variable', 'Variable:/:LENGTHX').Value =
5e-7
sim.createEntity('Variable', 'Variable:/:LENGTHY').Value =
5e-7
sim.createEntity('Variable', 'Variable:/:LENGTHZ').Value =
5e-7
```

5. *Define species, type and initial molecule numbers.* Each compartment is initially made up of empty voxels. We sometimes need to refer to these empty voxels in the simulation (*see* **Note 5**). These empty voxels are called VACANT species in the model and each compartment must have one such species declared to represent its empty voxels. By default, all species are non-HD type unless we set the Name property of the species to HD, as we have done for E, S, ES and P below. We have also set the species A, E and S to have 1500, 100, and 1000 molecules initially.

```
sim.createEntity('Variable', 'Variable:/:VACANT')
sim.createEntity('Variable', 'Variable:/:A').Value = 1500
sim.createEntity('Variable', 'Variable:/:B').Value = 0
sim.createEntity('Variable', 'Variable:/:C').Value = 0
v = sim.createEntity('Variable', 'Variable:/:E')
v.Value = 100
v.Name = 'HD'
v = sim.createEntity('Variable', 'Variable:/:S')
v.Value = 1000
v.Name = 'HD'
v = sim.createEntity('Variable', 'Variable:/:ES')
v.Value = 0
v.Name = 'HD'
```

```
v = sim.createEntity('Variable', 'Variable:/:P')
v.Value = 0
v.Name = 'HD'
```

6. *Populate non-HD species initialized with nonzero molecules.* In the model, only A is a non-HD species that has a nonzero initial number of molecules. We need to specify how to populate these explicitly represented molecules in the compartment using MoleculePopulateProcess. By default, the process will populate the initial 1500 molecules of A randomly in the compartment (*see* **Note 6** for alternative ways to populate). In the first line below, we have created a MoleculePopulateProcess object called 'Process:/:pop'. In the second line, we have connected the species A to the process by adding the reference of the variable to the process' VariableReferenceList. The first field (denoted here as '_') specifies a name of the variable reference that will be used to identify it locally in the process. MoleculePopulateProcess does not have any predefined variable reference name to identify connected variables, so we have just given an empty name field, '_'. The second field specifies the path, ':/:' and identity, A of the variable, which we have written here as 'Variable:/:A'. More details on how to connect variables to processes are provided in the E-Cell System manual available at https://ecell3.readthedocs.io/en/latest/modeling.html.

```
p = sim.createEntity('MoleculePopulateProcess', 'Process:/
:pop')
p.VariableReferenceList = [['_', 'Variable:/:A']]
```

7. *Set diffusion coefficient of non-HD species.* The DiffusionProcess is the module that specifies the diffusion properties of a species. We diffuse all three non-HD species, A, B, and C in the compartment with a diffusion coefficient of 0.0005 $\mu m^2 s^{-1}$.

```
d = sim.createEntity('DiffusionProcess', 'Process:/:d1')
d.VariableReferenceList = [['_', 'Variable:/:A']]
d.D = 5e-16
d = sim.createEntity('DiffusionProcess', 'Process:/:d2')
d.VariableReferenceList = [['_', 'Variable:/:B']]
d.D = 5e-16
d = sim.createEntity('DiffusionProcess', 'Process:/:d3')
d.VariableReferenceList = [['_', 'Variable:/:C']]
d.D = 5e-16
```

8. *Define the reactions.* The three deterministic reactions involving HD species are performed by MassActionProcess. We assign the ODEStepper to the reaction module by setting the StepperID to DE. SpatiocyteNextReactionProcess executes the stochastic reaction that generates B when P and A react. DiffusionInfluencedReactionProcess performs the bimolecular reaction between the two diffusing non-HD species, A and B, to produce C

(*see* **Note 7**). For the first mass action reaction E + S → ES, we need to connect the species E, S, and ES to a MassActionProcess object. As described in **step 6**, we connect them by adding the references of the variables into the process' VariableReference-List. Note that each new line of the VariableReferenceList with the operator '=' below does not overwrite the reference given in the previous line but adds the new reference to the existing list. Unlike in **step 6**, in the second field of the variable reference, we have used the relative path (w.r.t. the compartment path) of the variable, ':.:' instead of the absolute path, ':/:'. Relative path is useful when we want to skip updating the paths of the variables when we change the name of the compartment. The third field denotes whether the variable is a substrate ('-1') or a product ('1') of the reaction. The remaining reactions follow the same conventions to connect the species to the corresponding processes.

```
# E + S -> ES
r = sim.createEntity('MassActionProcess', 'Process:/:r1')
r.StepperID = 'DE'
r.VariableReferenceList = [['_', 'Variable:.:E','-1']]
r.VariableReferenceList = [['_', 'Variable:.:S','-1']]
r.VariableReferenceList = [['_', 'Variable:.:ES','1']]
r.k = 1e-22
# ES -> E + S
r = sim.createEntity('MassActionProcess', 'Process:/:r2')
r.StepperID = 'DE'
r.VariableReferenceList = [['_', 'Variable:.:ES', '-1']]
r.VariableReferenceList = [['_', 'Variable:.:E', '1']]
r.VariableReferenceList = [['_', 'Variable:.:S', '1']]
r.k = 1e-1
# ES -> E + P
r = sim.createEntity('MassActionProcess', 'Process:/:r3')
r.StepperID = 'DE'
r.VariableReferenceList = [['_', 'Variable:.:ES', '-1']]
r.VariableReferenceList = [['_', 'Variable:.:E', '1']]
r.VariableReferenceList = [['_', 'Variable:.:P', '1']]
r.k = 1e-1
# P + A -> B
r = sim.createEntity('SpatiocyteNextReactionProcess',
'Process:/:r4')
r.VariableReferenceList = [['_', 'Variable:/:P', '-1']]
r.VariableReferenceList = [['_', 'Variable:/:A', '-1']]
r.VariableReferenceList = [['_', 'Variable:/:B', '1']]
r.k = 5e-24
# A + B -> C
r = sim.createEntity('DiffusionInfluencedReactionProcess',
'Process:/:r5')
```

```
r.VariableReferenceList = [['_', 'Variable:/:A', '-1']]
r.VariableReferenceList = [['_', 'Variable:/:B', '-1']]
r.VariableReferenceList = [['_', 'Variable:/:C', '1']]
r.k = 5e-24
```

9. *Specify data loggers and the simulation time.* Below we use VisualizationLogProcess to log the coordinates of A, B, and C in lattice every 0.1 s in a binary format log file called VisualLog.dat (*see* **Note 8**). The Spatiocyte Visualizer can load the log file to display the 3D position of the molecules in time, while the simulation is running or after it has ended. We set the IteratingLogProcess to record each species' molecule number every 0.01 s from the beginning of the simulation until 99 s in a csv format file, IterateLog.csv (*see* **Note 9**). Finally, we can tell the simulator how long to run the model. Here, we set it to run for 100 s.

```
l = sim.createEntity('VisualizationLogProcess', 'Process:/
:l1')
l.VariableReferenceList = [['_', 'Variable:/:A']]
l.VariableReferenceList = [['_', 'Variable:/:B']]
l.VariableReferenceList = [['_', 'Variable:/:C']]
l.LogInterval = 1e-1
l = sim.createEntity('IteratingLogProcess', 'Process:/:
l2')
l.VariableReferenceList = [['_', 'Variable:/:A']]
l.VariableReferenceList = [['_', 'Variable:/:B']]
l.VariableReferenceList = [['_', 'Variable:/:C']]
l.VariableReferenceList = [['_', 'Variable:.:E']]
l.VariableReferenceList = [['_', 'Variable:.:S']]
l.VariableReferenceList = [['_', 'Variable:.:ES']]
l.VariableReferenceList = [['_', 'Variable:.:P']]
l.LogInterval = 1e-2
l.LogEnd = 99
run(100)
```

3.2 Run the Model

After successfully installing Spatiocyte (*see* **Note 10**), we can simulate the multi-algorithm model in a terminal by issuing

```
$ ecell3-session ode-snrp-particle.py
```

The simulator will run for 100 s and terminate. In the current working directory, it would have saved two log files, VisualLog.dat and IterateLog.csv.

3.3 Display Simulation Results

Finally, we can visualize the data logged by the two loggers with the following steps.

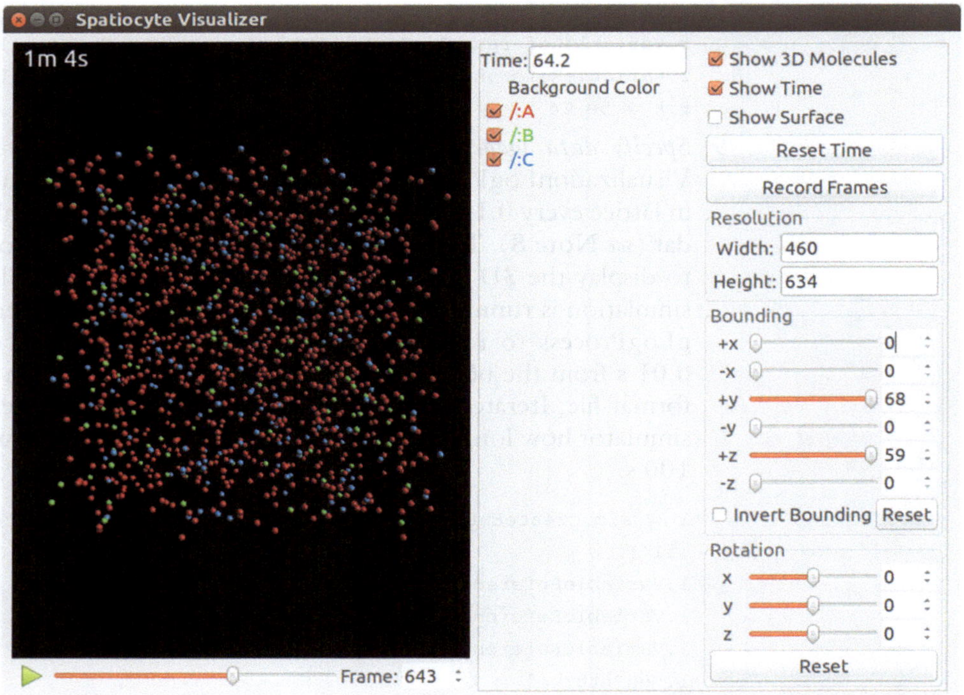

Fig. 2 The graphical user interface of Spatiocyte Visualizer

1. *View diffusing molecules.* Even while the simulation is running we can view the dynamics of the diffusing molecules with Spatiocyte Visualizer (Fig. 2) by issuing

```
$ spatiocyte VisualLog.dat
```

   in the working directory. The visualizer will load VisualLog.dat and display the molecule positions of non-HD species, A, B, and C, as displayed in Fig. 3. The shortcut keys to control the visualizer are provided in the Spatiocyte guide [26]. For example, the right arrow key will advance the time forward whereas the left arrow key, backward. Pressing the space bar key will pause or resume the time advancement.

2. *View time course profiles.* From IterateLog.csv, we can plot the time course profiles of the logged species using a helper Python script called plotIterateLog.py, which is included in the Spatiocyte examples/plot directory. Copying the file into the working directory and issuing the command below will display the profiles, as shown in Fig. 4.

```
$ python plotIterateLog.py
```

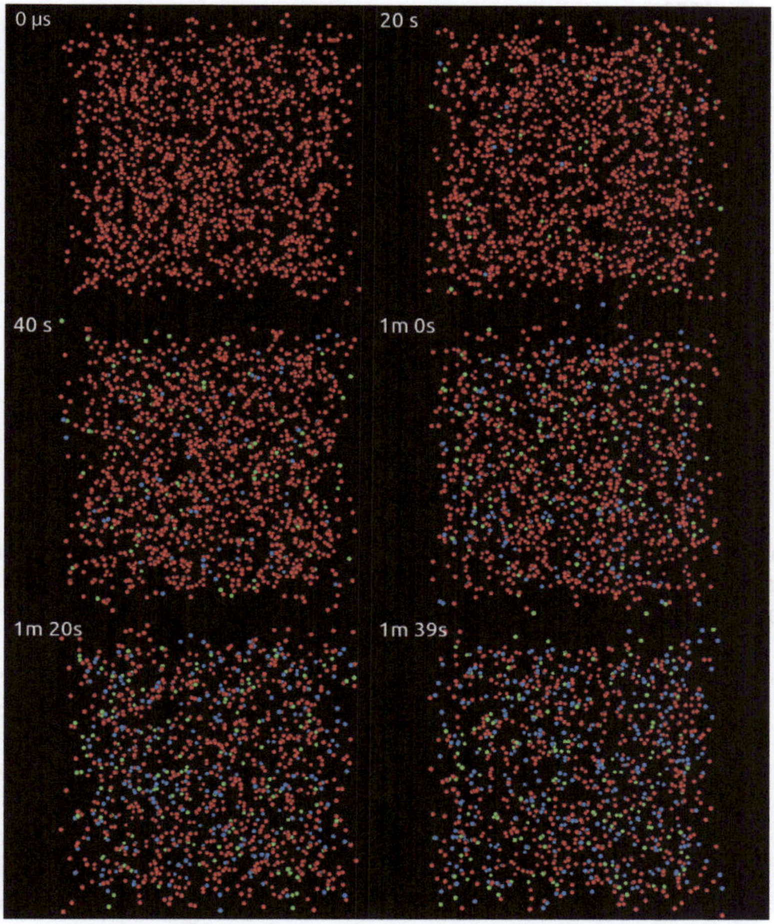

Fig. 3 Simulation snapshots of the multi-algorithm model. Initially, all 1500 molecules of A (*red*) are populated randomly in cubic space. As time advances, more B (*green*) and C (*blue*) molecules start to appear while A decreases as a result of the multi-algorithm reactions

4 Case Studies

We have previously used Spatiocyte to model *Escherichia coli* division site regulators, MinD and MinE proteins that periodically cycle the poles of the rod-shaped bacterium [24]. Our model is the first to corroborate the prediction that MinE can bind to the membrane independently using its membrane domain [27, 28] after it is recruited from the cytoplasm by MinD. The model also first predicted that independently membrane-bound MinE can rebind with other MinD's on the membrane. These predictions were later supported experimentally [29, 30]. Recently, we built a multi-algorithm simulation model of erythrocyte band 3 membrane cluster formation with Spatiocyte [31]. The model showed that strong

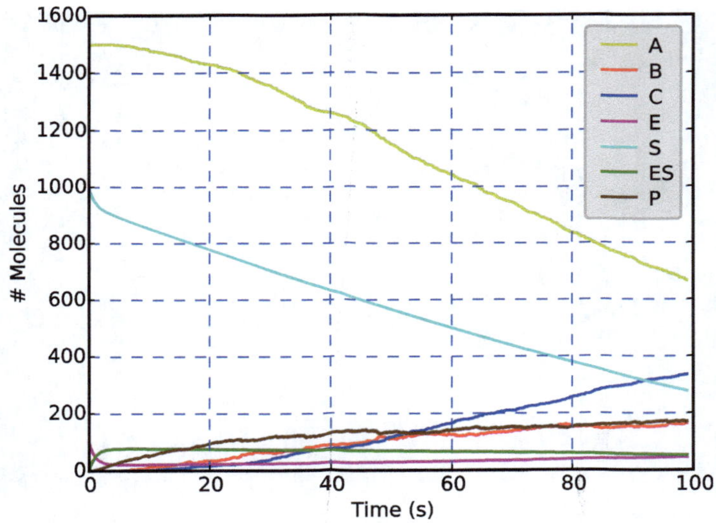

Fig. 4 Time course profiles of the multi-algorithm simulation. S, E, and ES show smooth lines over time because they are only involved in deterministic reactions. P, which is involved in both stochastic and deterministic reactions, displays noisy increase over time

affinity between the clustering molecules and irreversibly binding hemichromes aid the generation of oxidation induced clusters as observed in experiments. The simulated cluster size increased toward an irreversible state when oxidative stress is introduced repeatedly. The model also predicted that erythrocytes with deficient spectrin cytoskeletal filaments have more and larger band 3 clusters. In addition, together with our colleagues, we have recently developed a bioimaging simulation framework that produces simulated microscopy images from 3D molecule coordinates generated by particle simulators such as Spatiocyte [32]. The simulated images can be compared with actual microscopy images at the level of photon-counting units. We verified the bioimaging simulator by comparing simulated images of several in vitro and in vivo Spatiocyte models with experimentally obtained microscopy images.

Here, as another example of Spatiocyte application, we show that *E. coli* cell geometry can regulate MinD oscillation period, while the cell size controls the peak MinD concentration on the membrane. Previous works have shown that MinD dynamics can be regulated by the geometry [33–37] and topology [38] of the membrane. Varma and colleagues [33] used *E. coli* lacking penicillin binding proteins to produce branched cells with three poles (Y-shaped) to investigate the effects of the mutant cell geometry on MinD membrane dynamics. Cells having almost equal branch lengths displayed non-reversing clockwise or counterclockwise rotational MinD polar localization. In cells where two of its poles are closer to each other than the third pole, MinD cycled back and forth symmetrically between the two poles and the third pole.

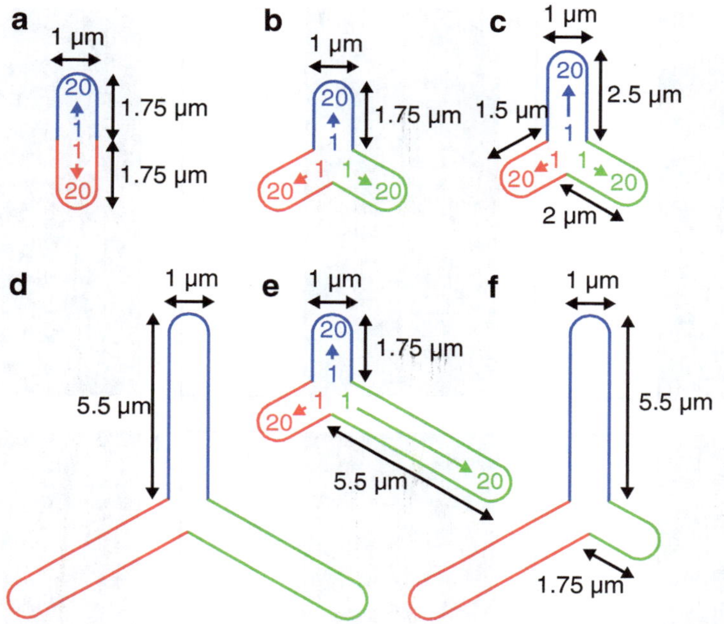

Fig. 5 Schematic representation of *E. coli* geometric configurations used in simulations. *Blue*, *red*, and *green borders* indicate the different branches of the cell. The range 1–20 in each branch represents the bin # in which the membrane concentration of MinD is calculated. The length of a branch is not stated if the length is already specified for another branch of the cell having the same length. (**a**) Wild type geometry; (**b–f**) Branched geometry with different branch lengths

Adapting our previously reported model [24], we investigated MinD dynamics in different geometric configurations of the branched cells as illustrated in Fig. 5, with fixed protein concentrations. Fig. 6 displays the corresponding kymographs of MinD simulation results. In cells with equal branch lengths of 1.75 μm, MinD showed symmetrical oscillation that occasionally switched poles randomly. In all other configurations, MinD produced stable symmetrical oscillations. Despite implementing such diverse geometric configurations of the branched cells, our model is unable to recapitulate the rotational oscillation as observed in the experiments. Further detailed simulations are necessary to identify the requirements of rotational MinD oscillation in branched cells. Nonetheless, our preliminary simulations indicate that the period of oscillation increases as the total length of the branches increases. Regardless of the cell geometries, the peak concentration of MinD on the membrane correlates with the total surface area or the volume of the cell. More simulations and analyses are required to reveal how the oscillation period and MinD membrane concentration are regulated by the branch lengths, cell volume, and membrane surface area.

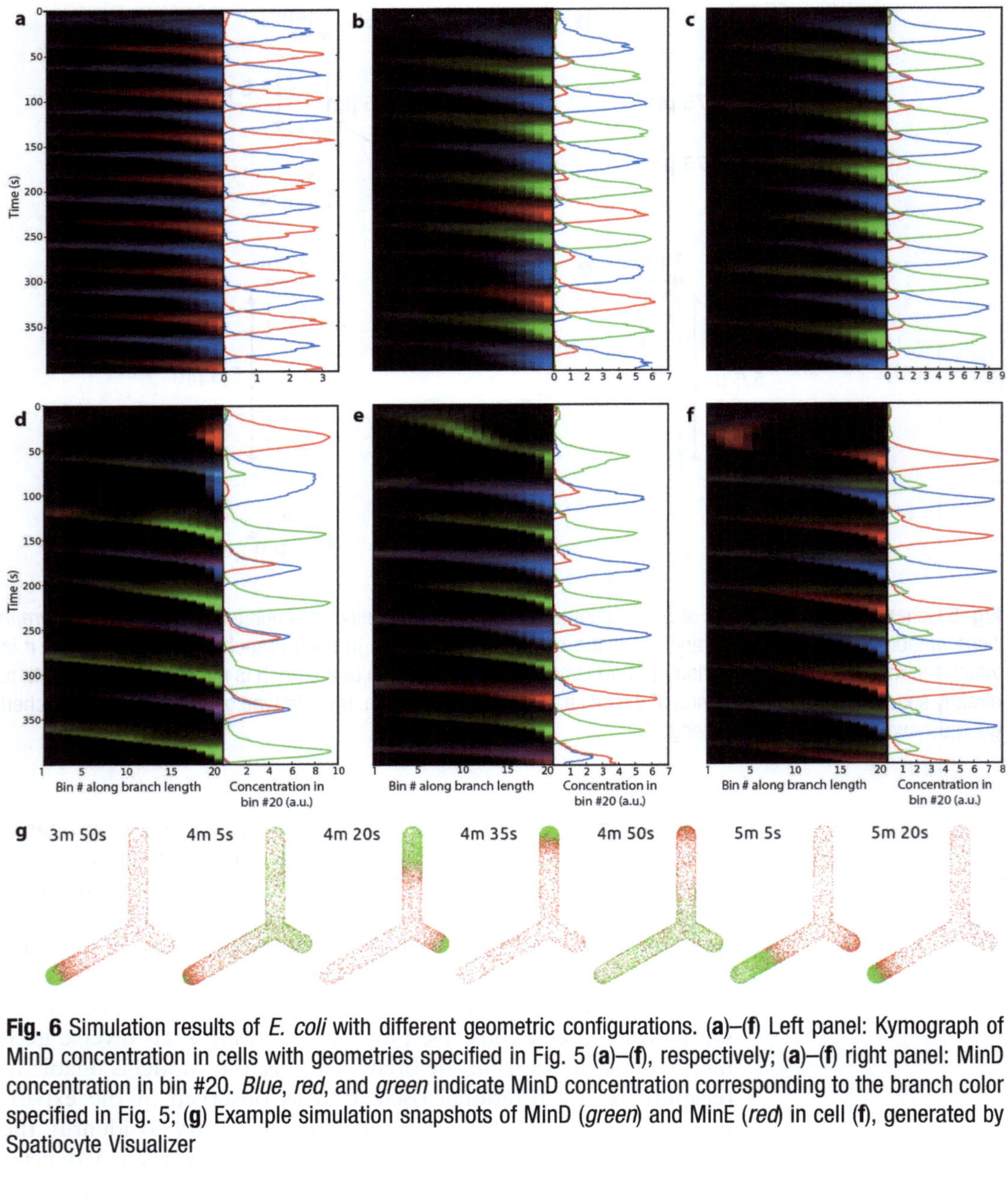

Fig. 6 Simulation results of *E. coli* with different geometric configurations. (**a**)–(**f**) Left panel: Kymograph of MinD concentration in cells with geometries specified in Fig. 5 (**a**)–(**f**), respectively; (**a**)–(**f**) right panel: MinD concentration in bin #20. *Blue*, *red*, and *green* indicate MinD concentration corresponding to the branch color specified in Fig. 5; (**g**) Example simulation snapshots of MinD (*green*) and MinE (*red*) in cell (**f**), generated by Spatiocyte Visualizer

5 Notes

1. *Modeling language.* Spatiocyte models can be written either in Python or C++. With Python, we can simulate the model without compiling it into an executable, which would take up additional time and effort. It is also easier with Python to perform multiple iterations of a model and introduce

conditions when running the simulation because it is a scripting language. C++ models on other hand permit more flexibility and are useful when we want to optimize the compiled executable for a specific CPU for faster run times.

2. *VoxelRadius value.* For better simulation accuracy, the value of VoxelRadius should be close to the hydrodynamic radius of the diffusing species [24]. However, the simulation would consume more computation time when the VoxelRadius is small because of the shorter simulation time steps required when performing smaller diffusion steps over the voxels. The memory usage also increases linearly with the number of voxels. In a 64-bit system, each voxel typically takes up 108 bytes of memory. The number of voxels with radius, r, in a volume, V, is given by $V/(4r^3 2^{0.5})$. Therefore, in the initial stages of modeling, we usually first perform quick simulations with larger voxels and attempt to recapitulate experimentally observed phenotypes by modifying reactions and other unknown model parameter values. As the simulation phenotypes start to agree with observations, we gradually reduce the size of the voxels to the hydrodynamic radius of diffusing species.
3. *MaxStepInterval of ODEStepper.* ODEStepper executes mass action reactions at varying step intervals. To allow fast simulations, it dynamically increases the step interval when accuracy would not be compromised for the reactions. However, SpatiocyteStepper typically performs diffusion-influenced-reaction and next-reaction at very short intervals because of the short diffusion time steps. To ensure that the molecule number of the species in the mass action reactions are valid at these short intervals when they are accessed by SpatiocyteStepper reactions, we set the MaxStepInterval of the ODEStepper to a small value.
4. *StepperID inheritance.* The StepperID for all modules in a compartment such as DiffusionProcess, MassActionProcess, and SpatiocyteNextReactionProcess is inherited from the compartment's StepperID. Since all modules except MassActionProcess are executed in event-driven manner by SpatiocyteStepper, we set the root compartment's StepperID to the SpatiocyteStepper ID. In each MassActionProcess module, we can directly set its StepperID to the ODEStepper.
5. *Reactions involving VACANT species.* In some reactions, we need to specify the VACANT species of the compartment as a reactant. For example, in the diffusion-influenced membrane association reaction, where a cytosolic A binds to the membrane to form Am, the VACANT voxels of the membrane compartment are one of the reactants of the second-order reaction: A + membrane:VACANT – > Am.

6. *Populating molecules in a compartment.* Non-HD molecules are by default randomly populated throughout the compartment of the species with uniform distribution by MoleculePopulateProcess. We can also set a specific range to populate along each dimension of the compartment by setting the Origin[X, Y, Z] and Uniform[Length, Width, Height] options of the process. Molecules can also be populated along the length of the compartment divided into a given number of bins with different occupancy fractions using the LengthBinFractions array option. It specifies the number of bins and the population fraction of molecules over the total available vacant voxels in each bin.
7. *Reaction module selection.* We use DiffusionInfluencedReactionProcess only for second-order reactions where both reactants are diffusing non-HD species. If all the reactants of first- or second-order reaction are HD, we can use MassActionProcess. For all first-order reactions we can also use SpatiocyteNextReactionProcess. We implement SpatiocyteNextReactionProcess for all second-order reactions when either (or both) of the reactants is HD.
8. *VisualizationLogProcess default log interval.* If we do not specify the LogInterval value of VisualizationLogProcess, the logger will log the coordinates of its listed diffusing species at all SpatiocyteStepper time steps. This option is useful when we want to detect the exact time when a molecule changes its state in space.
9. *Log molecule coordinates in csv format.* Spatiocyte also comes with another logger module called CoordinateLogProcess that saves the coordinates of non-HD molecules at defined intervals in csv format. The coordinate data is useful for the user to perform custom detailed analysis of the simulation. The log file can also be read by a helper Python plotting script called plotCoordinateLog.py, included in the Spatiocyte examples directory.
10. *Verifying Spatiocyte installation.* To test if the Spatiocyte installation is successful, issue the following command in a terminal: `$ ecell3-session`. The above command will start the Python command line interface of Spatiocyte. If for some reason, the interface does not come up, the error message can be posted to the Spatiocyte Users forum at https://groups.google.com/forum/?hl=en#!forum/spatiocyte-users for help.

Acknowledgment

We thank Masaki Watabe, Hanae Shimo, and Kaizu Kazunari for discussions that led to the improvement of Spatiocyte usage. We also appreciate Kozo Nishida for Spatiocyte software packaging, installation, and documentation assistance.

References

1. Fange D, Elf J (2006) Noise-induced min phenotypes in *E. coli*. *PLoS Comput Biol 2*(6):e80. doi:10.1371/journal.pcbi.0020080
2. Hecht I, Kessler DA, Levine H (2010) Transient localized patterns in noise-driven reaction-diffusion systems. Phys Rev Lett 104 (15):158301. doi:10.1103/PhysRevLett.104.158301
3. Burrage K, Burrage PM, Marquez-lago T, Nicolau DV (2011) Stochastic simulation for spatial modelling of dynamic processes in a living cell. In: Koeppl H, Setti G, di Bernardo M, Densmore D (eds) Design and analysis of biomolecular circuits: engineering approaches to systems and synthetic biology. Springer, New York, NY, pp 43–62. doi:10.1007/978-1-4419-6766-4
4. Klann M, Koeppl H (2012) Spatial simulations in systems biology: from molecules to cells. Int J Mol Sci 13(6):7798–7827. doi:10.3390/ijms13067798
5. Schöneberg J, Ullrich A, Noé F (2014) Simulation tools for particle-based reaction-diffusion dynamics in continuous space. BMC Biophys 7(1):11. doi:10.1186/s13628-014-0011-5
6. Karr JR, Takahashi K, Funahashi A (2015) The principles of whole-cell modeling. Curr Opin Microbiol 27:18–24. doi:10.1016/j.mib.2015.06.004
7. Kerr RA, Bartol TM, Kaminsky B, Dittrich M, Chang J-CJ, Baden SB, Sejnowski TJ, Stiles JR (2008) Fast Monte Carlo simulation methods for biological reaction-diffusion Systems in Solution and on surfaces. SIAM J Sci Comput 30(6):3126–3149. doi:10.1137/070692017
8. Fange D, Mahmutovic A, Elf J (2012) MesoRD 1.0: Stochastic reaction-diffusion simulations in the microscopic limit. *Bioinformatics* 28:1–3. doi:10.1093/bioinformatics/bts584
9. Angermann, B. R., Klauschen, F., Garcia, A. D., Prustel, T., Zhang, F., Germain, R. N., & Meier-Schellersheim, M. (2012). Computational modeling of cellular signaling processes embedded into dynamic spatial contexts. *Nat Methods*, (2011), 1–10. doi:10.1038/nmeth.1861
10. Drawert B, Engblom S, Hellander A (2012) URDME : a modular framework for stochastic simulation of reaction-transport processes in complex geometries. BMC Syst Biol 6 (76):1–17. doi:10.1186/1752-0509-6-76
11. Hepburn I, Chen W, Wils S, De Schutter E (2012) STEPS: efficient simulation of stochastic reaction-diffusion models in realistic morphologies. BMC Syst Biol 6(1):36. doi:10.1186/1752-0509-6-36
12. Roberts E, Stone JE, Luthey-Schulten Z (2012) Lattice microbes: high-performance stochastic simulation method for the reaction-diffusion master equation. J Comput Chem. doi:10.1002/jcc.23130
13. Andrews SS, Addy NJ, Brent R, Arkin AP (2010) Detailed simulations of cell biology with Smoldyn 2.1. *PLoS Comput Biol 6*(3): e1000705. doi:10.1371/journal.pcbi.1000705
14. Byrne MJ, Waxham MN, Kubota Y (2010) Cellular dynamic simulator: an event driven molecular simulation environment for cellular physiology. Neuroinformatics 8(2):63–82. doi:10.1007/s12021-010-9066-x
15. Takahashi K, Tanase-Nicola S, ten Wolde PR (2010) Spatio-temporal correlations can drastically change the response of a MAPK pathway. Proc Natl Acad Sci U S A 107 (6):2473–2478. doi:10.1073/pnas.0906885107
16. Tolle DP, Le Novere N (2010) Meredys, a multi-compartment reaction-diffusion simulator using multistate realistic molecular complexes. BMC Syst Biol 4(1):24. doi:10.1186/1752-0509-4-24
17. Schöneberg J, Noé F (2013) ReaDDy—a software for particle-based reaction-diffusion dynamics in crowded cellular environments. PLoS One 8(9):e74261. doi:10.1371/journal.pone.0074261
18. Karamitros, M., Luan, S., Bernal, M. A., Allison, J., Baldacchino, G., Davidkova, M., Z. Francis, W. Friedland, V. Ivantchenko, A. Ivantchenko, A. Mantero, P. Nieminem, G. Santin, H.N. Tran, V. Stepan, Incerti, S. (2014). Diffusion-controlled reactions

modeling in Geant4-DNA. J Comput Phys, 274, 841–882. doi:10.1016/j.jcp.2014.06.011

19. Michalski PJ, Loew LM (2016) SpringSaLaD: a spatial, particle-based biochemical simulation platform with excluded volume. Biophys J 110 (3):523–529. http://doi.org/10.1016/j.bpj.2015.12.026
20. Hellander A, Hellander S, Lötstedt P (2012) Coupled mesoscopic and microscopic simulation of stochastic reaction-diffusion processes in mixed dimensions. Multiscale Model Simul 10(2):585–611. doi:10.1137/110832148
21. Klann M, Ganguly A, Koeppl H (2012) Hybrid spatial Gillespie and particle tracking simulation. Bioinformatics 28(18):i549–i555. doi:10.1093/bioinformatics/bts384
22. Robinson M, Andrews SS, Erban R (2015) Multiscale reaction-diffusion simulations with Smoldyn. Bioinformatics 31(14):2406–2408. http://doi.org/10.1093/bioinformatics/btv149
23. Arjunan SNV, Kaizu K, Takahashi K. Spatiocyte: a stochastic particle simulator for filament, membrane and cytosolic reaction-diffusion processes. In preparation.
24. Arjunan SNV, Tomita M (2010) A new multi-compartmental reaction-diffusion modeling method links transient membrane attachment of *E. coli* MinE to E-ring formation. Syst Synth Biol 4(1):35–53. doi:10.1007/s11693-009-9047-2
25. Gibson MA, Bruck J (2000) Efficient exact stochastic simulation of chemical systems with many species and many channels. J Phys Chem A 104(9):1876–1889. doi:10.1021/jp993732q
26. Arjunan SNV (2013) A guide to modeling reaction-diffusion of molecules with the E-cell system. In: Arjunan SNV, Tomita M, Dhar PK (eds) E-cell system: basic concepts and applications. Springer Science & Business Media, New York, NY
27. King GF, Rowland SL, Pan B, Mackay JP, Mullen GP, Rothfield LI (1999) The dimerization and topological specificity functions of MinE reside in a structurally autonomous C-terminal domain. Mol Microbiol 31(4):1161–1169. doi:10.1046/j.1365-2958.1999.01256.x
28. Ma L-Y, King G, Rothfield L (2003) Mapping the MinE site involved in interaction with the MinD division site selection protein of *Escherichia coli*. J Bacteriol 185(16):4948–4955. doi:10.1128/JB.185.16.4948-4955.2003
29. Loose M, Fischer-Friedrich E, Herold C, Kruse K, Schwille P (2011) Min protein patterns emerge from rapid rebinding and membrane interaction of MinE. Nat Struct Mol Biol 18 (5):577–583. doi:10.1038/nsmb.2037
30. Park K-T, Wu W, Battaile KP, Lovell S, Holyoak T, Lutkenhaus J (2011) The Min oscillator uses MinD-dependent conformational changes in MinE to spatially regulate cytokinesis. Cell 146(3):396–407. doi:10.1016/j.cell.2011.06.042
31. Shimo H, Arjunan SNV, Machiyama H, Nishino T, Suematsu M, Fujita H, Tomita M, Takahashi K (2015) Particle simulation of oxidation induced band 3 clustering in human erythrocytes. PLoS Comput Biol 11(6): e1004210. doi:10.1371/journal.pcbi.1004210
32. Watabe M, Arjunan SNV, Fukushima S, Iwamoto K, Kozuka J, Matsuoka S, Shindo Y, Ueda M, Takahashi K (2015) A computational framework for bioimaging simulation. PLoS One 10(7):e0130089. doi:10.1371/journal.pone.0130089
33. Varma A, Huang KC, Young KD (2008) The Min system as a general cell geometry detection mechanism: branch lengths in Y-shaped *Escherichia coli* cells affect Min oscillation patterns and division dynamics. J Bacteriol 190 (6):2106–2117. doi:10.1128/JB.00720-07
34. Schweizer J, Loose M, Bonny M, Kruse K, Monch I, Schwille P (2012) Geometry sensing by self-organized protein patterns. Proc Natl Acad Sci 109(38):15283–15288. doi:10.1073/pnas.1206953109
35. Halatek J, Frey E (2014) Effective 2D model does not account for geometry sensing by self-organized proteins patterns. Proc Natl Acad Sci 111(18):E1817–E1817. doi:10.1073/pnas.1220971111
36. Wu F, van Schie BGC, Keymer JE, Dekker C (2015) Symmetry and scale orient Min protein patterns in shaped bacterial sculptures. Nat Nanotechnol 10(8):719–726. doi:10.1038/nnano.2015.126
37. Zieske K, Schwille P (2015) Reconstituting geometry-modulated protein patterns in membrane compartments. Methods Cell Biol 128:149–163. doi:10.1016/bs.mcb.2015.02.006
38. Zieske K, Schweizer J, Schwille P (2014) Surface topology assisted alignment of Min protein waves. FEBS Lett 588(15):2545–2549. doi:10.1016/j.febslet.2014.06.026

INDEX

Daisuke Kihara (ed.), *Protein Function Prediction: Methods and Protocols*, Methods in Molecular Biology, vol. 1611, DOI 10.1007/978-1-4939-7015-5, © Springer Science+Business Media LLC 2017

MIX
Papier aus verantwortungsvollen Quellen
Paper from responsible sources
FSC® C105338

If you have any concerns about our products,
you can contact us on
ProductSafety@springernature.com

In case Publisher is established outside the EU,
the EU authorized representative is:
Springer Nature Customer Service Center GmbH
Europaplatz 3, 69115 Heidelberg, Germany

Printed by Libri Plureos GmbH
in Hamburg, Germany